TWENTY FIRST CENTURY
SCIⓔNCE

D1344657

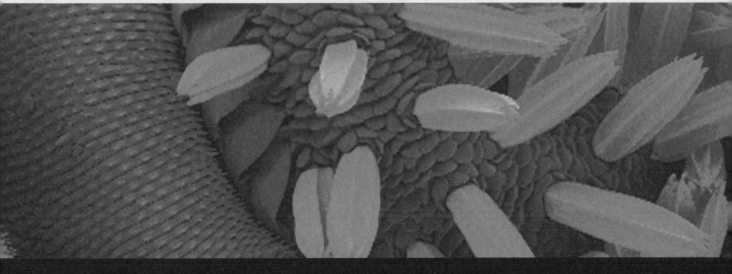

Project Directors

Angela Hall   Emma Palmer

Robin Millar   Mary Whitehouse

Editors

Angela Hall

Emma Palmer

Mary Whitehouse

Authors

Colin Bell	Andrew Hunt	Carol Tear
Peter Campbell	Carol Levick	Carol Usher
Byron Dawson	Caroline Shearer	Vicky Wong

THE UNIVERSITY *of York*

THE SALTERS' INSTITUTE

Nuffield Foundation

OCR RECOGNISING ACHIEVEMENT   OXFORD UNIVERSITY PRESS
Official Publisher Partnership

Contents

Topic 1: Peak performance

A The skeletal system

① Living skeleton

Vertebrates need a skeleton for support and movement.

a Fill in the spaces to complete the following definitions. Use the words in the box.

backbone	support	framework	joints	movement

A vertebrate is an animal with a ....

An internal skeleton is a living structure that has ....

for movement. The skeleton is also used for ....

b Describe some features of the human skeleton that help it to:

i provide support ....

....

....

ii allow movement ....

....

....

Joints and movement

① **Joints**

Movement of a joint depends on all its parts functioning correctly.

a Draw a straight line from each key word to its correct meaning.

cartilage	fluid lubricating and nourishing a joint
joint	tissue joining bones together
ligament	tissue joining muscle to bone
synovial fluid	smooth, shock-absorbing tissue protecting bones
tendon	where two or more bones meet

b Draw and label a joint between two bones showing bones, ligaments, cartilage, and synovial fluid.

c Specific properties of joint tissues enable them to function effectively. Complete the table to:

• describe the function of each tissue

• describe some specific properties of each tissue.

Tissue	Function	Specific properties
cartilage		
ligaments		
tendons		

d i Draw two muscles that move the elbow joint on the diagram of the bones of the arm. You must show which bones they connect to.

Label the muscles A and B.

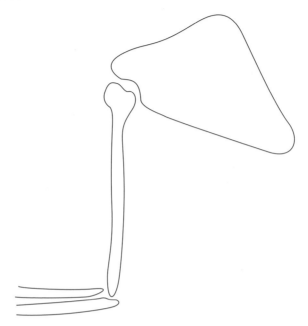

ii Explain what happens to each muscle when the lower arm is raised.

A .... B ....

iii Explain what happens to each muscle when the lower arm is lowered.

A .... B ....

iv These muscles are an antagonistic pair. Explain what this term means.

....

....

v Identify an antagonistic pair of muscles from another part of the body.

....

C Starting an exercise regime

① Training programmes

Physical training programmes can improve fitness and aid recovery from illness or injury.

a Make a list of the sort of medical and lifestyle information that a trainer or medical professional would need to know about a client before recommending a training programme.

....

....

....

b Explain how a trainer working with an athlete would use their medical and lifestyle history.

....

....

c Why does a fitness trainer need to know if their client starts a new course of medication?

....

d Describe an example of how the progress of a sportsman or sportswoman is monitored over a season of training. Give examples of what measurements would be taken and recorded.

....

....

....

....

e Explain how the following changes help to provide the extra energy needed by muscle cells during exercise.

   i The breathing rate increases.

....

....

....

   ii The heart rate increases.

....

....

....

(2) Assessment, modification, and follow up

a Treatments often have side-effects, which have to be weighed up against the benefits.

Add examples of possible benefits and side-effects of the treatments suggested in the table.

Suggested treatment	Possible benefits	Possible side-effects
A a gym workout programme for an overweight person (target: weight in the 'normal' range)		
B a walking programme for a patient with heart problems (target: improved cardiovascular fitness)		

b Choose one of the examples (A or B) in the table, and suggest why the programme might need to be modified before it has been completed.

....

....

c For example A from the table, suggest another way of exercising to achieve the target.

....

....

d Suggest how the fitness and progress of the person in example A from the table could be monitored.

....

e Assessment of progress needs to take into account the accuracy of the monitoring technique of the data obtained.

Explain what problems there might be with data obtained in these situations.

- **Treatment A** (in the table) Occasionally measuring his/her weight at home using bathroom scales.

....

- **Treatment B** Having blood pressure and pulse measured weekly by a visiting nurse who calls at a different time each week.

....

f Explain why 'normal' measurements for heart rate and blood pressure cover a range of values.

....

....

③ Interpreting data

When you exercise, your heart works harder. Jim and Kirsty have been training for one month by running on a treadmill at the gym.
The table below shows measurements for Jim and Kirsty's heart rate and blood pressure before exercise and after 20 minutes of running on a treadmill.

	Jim		Kirsty	
	Before exercise	After exercise	Before exercise	After exercise
Body mass (kg)	91		65	
Height (cm)	173		180	
BMI				
Pulse rate (b.p.m)	73	170	61	150
Blood pressure (mmHg)	110/80	130/80	100/70	105/72
Recovery rate (time taken for pulse to return to reading before exercise) (min)		20		7

a Calculate the BMI for Jim and Kirsty and write the values in the table.

b Who is the fittest person?

c Give three reasons for your answer to part **b**.

1

2

3

d Both Jim and Kirsty have been on a training programme at the gym for a month. How would you adjust their programme after these tests? Give reasons for your answer.

Jim

Reason

Kirsty

Reason

e What other measurement could you take?

f How would the measurement you named in part **e** affect a trainer's decision about the exercise regime?

Fit for purpose?

① Accuracy and error

Measuring equipment should be accurate and give repeatable results.

The next day Jim was monitored again. Here are his results.

	Jim (Day 1)		Jim (Day 2)	
	Before exercise	After exercise	Before exercise	After exercise
Body mass (kg)	91		89	
Height (cm)	173		173	
BMI				
Pulse rate (b.p.m)	73	170	69	159
Blood pressure (mmHg)	110/80	130/80	130/80	150/90
Recovery rate (time taken for pulse to return to reading before exercise) (min)		20		19

a Make a list of the differences and say if this is significant or not.

	Difference	Significant?
Body mass		
Pulse rate		
Blood pressure		
Recovery rate		

b How could the gym make their monitoring as accurate as possible?

....

....

....

Sports injuries

1 Speeding up healing

The correct treatment of skeletal–muscular injuries can help the healing process.

a Complete the table to show the correct name of each injury. Use names from the box.

| dislocation | sprain | torn ligament | torn tendon |

Description	Injury
tissue damage that results in an unstable joint	
tissue damage that results in loss of a certain movement	
an overstretched ligament	
displacement of a bone from its normal position in a joint	

b The most common sporting injury is a sprain. Complete the descriptions of the three main symptoms of a sprain.

1 Appearance (shape):

2 Appearance (colour):

3

c The immediate treatment for sprains is RICE. List what each letter stands for, and for each letter describe what you would do to help someone who had sprained an ankle.

d Describe the sort of treatment that might follow a suitable period of RICE treatment.

① **Physiotherapy treatment**

Physiotherapy is an important part of treating skeleton–muscular injuries.

a If you have a skeletal–muscular injury you may need the help of a physiotherapist.

   i At what stage would a physiotherapist be involved in your treatment?

     ....

   ii Describe how a physiotherapist would help to treat your injury.

     ....

     ....

     ....

b Read the list of exercises recommended to someone recovering from a joint injury, then answer the questions below. Include references to muscles, tendons, and ligaments in your answers.

> **Exercise 1**
> Lie with your leg out straight. Tense up your thigh muscles, push your knee down, and try to raise your heel. Hold for a few seconds.
>
> **Exercise 2**
> Place a rolled up towel under your knee, keep your knee on the roll, and lift your heel. Try to get your knee completely straight.
>
> **Exercise 3**
> Bend your knee as far as it can easily go. Hold for a few seconds then straighten and repeat.
>
> **Exercise 4**
> Lie on your front. Keep your thigh down and bend your knee as far as you easily can.

   i What do you think is the main purpose of these exercises?

     ....

   ii Why do you think all the exercises are done lying down?

     ....

Topic 2

Blood

① The circulatory system

This is made up of the heart, arteries, capillaries, veins, and blood. The sentences below describe the system for moving blood around the body. Number the sentences in the right order. One has been done for you.

| 1 | The heart pumps blood to the lungs to collect oxygen and remove carbon dioxide. |

☐ The heart pumps blood around the body.

☐ Deoxygenated blood returns to the heart through veins.

☐ Oxygenated blood returns to the heart.

☐ In capillaries, oxygen and glucose diffuse into cells and waste products (carbon dioxide and urea) diffuse into the blood.

② Blood and its components

Blood is a tissue. It contains different types of cells.

a Draw a straight line from each blood component to its correct function.

Blood components

red blood cells

white blood cells

platelets

plasma

Function

to transport water, solutes, and heat

to clot blood at injury sites

to fight infection

to transport oxygen

b Draw a red blood cell and label with notes to explain how it is adapted for its function. Include the following words.

haemoglobin   membrane   biconcave

① **Double circulation**

a The heart pumps blood around the body.

- On the diagram add arrows between the boxes to show the double circulation of the blood around the body.

- Label the points where there is a capillary network.

lungs	heart	rest of body

b Explain why a double circulation is important for oxygenated blood to reach all parts of the body.

....

....

c Complete the table describing different parts of the circulatory system.

Circulatory system structure	Description
	Heart chamber that receives blood from all parts of the body except the lungs
	pumps blood to the lungs
	receives blood from the lungs
	pumps blood to all parts of the body except the lungs
vena cava	
	transports blood from the heart to the body
pulmonary vein	
pulmonary artery	
	supplies blood to the heart muscle

d In the above table:

- colour **red** the circulatory system structures that are filled with *oxygenated* blood

- colour **blue** the structures that are filled with *deoxygenated* blood

e The left ventricle has a thicker muscle wall than the right ventricle. How is this important in the working of the heart?

....

Valves and tissue fluid

① Valves

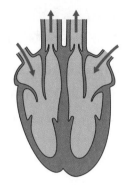

a The diagram on the right shows the valves in the heart.

   i Draw a ⟨ring⟩ around the positions of the four heart valves.

   ii Explain what the valves in the heart do.

....

....

b Complete the diagram to show how the valves in veins work.

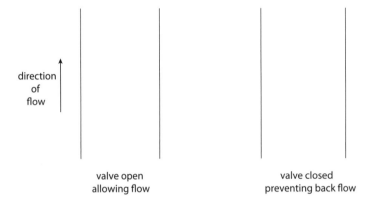

direction
of
flow

valve open
allowing flow

valve closed
preventing back flow

② Capillaries and tissue fluid

The diagram shows a capillary and the cells that surround it. Explain how nutrients
pass from the blood to the cells and how waste products pass from the cells to the blood.

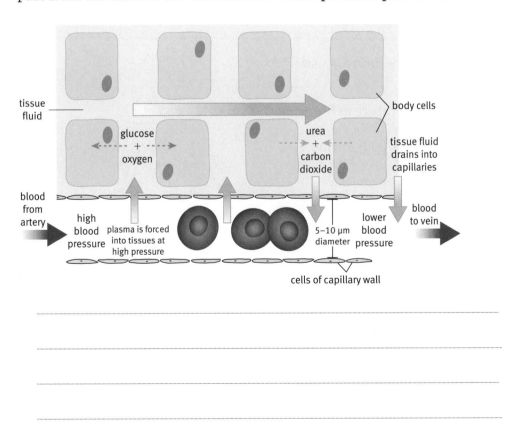

tissue
fluid

glucose
+
oxygen

urea
+
carbon
dioxide

body cells

tissue fluid
drains into
capillaries

blood
from
artery

high
blood
pressure

plasma is forced
into tissues at
high pressure

5–10 µm
diameter

lower
blood
pressure

blood
to vein

cells of capillary wall

....

....

....

....

A	**Getting hot, getting cold**

① **Balancing body temperature**

Animals gain heat from and lose heat to their environment.

a The diagram below shows the direction heat is moving. Decide which is hotter – the body or the environment. Colour the hot areas **red** and the cooler areas **blue**.

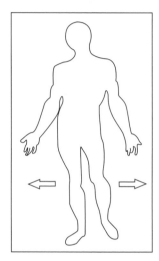

b Complete this sentence.

To keep a steady body temperature, the following must be true:

heat .... = heat ....

c Not all of your body is at the same temperature. Answer the questions using the diagram.

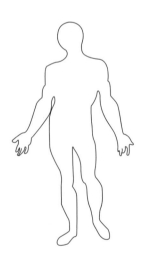

  i What is the normal core body temperature in humans?

   Think about the temperature of your body surface.

  ii Colour in **blue** the areas that would be coolest.

  iii Colour in **red** the areas that would be hottest.

  iv Why are these areas hotter?

  ....

  v How is heat transferred to other parts of the body?

  ....

d Sir Charles Blagden was able to stay in a room at about 127°C for eight minutes.

During that time how was his body maintaining a steady temperature?

_____

Sensing and control

① **Body temperature control systems**

This text describes how your body controls its temperature.
Read the text then answer the questions.

Receptors in your skin detect changes in the temperature of the air around you. Receptors in your brain detect changes in the temperature of your blood. This information is passed to the temperature control centre in the brain – called the hypothalamus.

The hypothalamus coordinates all the information from the receptors. It automatically triggers effectors in the body to respond to changes in body temperature. These effectors are sweat glands and muscles.

If body temperature rises too high:

- sweat glands increase production of sweat
- muscles in blood vessels supplying the skin relax

These responses work to lower body temperature.

If body temperature becomes too low:

- sweat production is reduced
- muscles in blood vessels supplying the skin contract
- skeletal muscles contract rapidly, causing shivering

These responses work to raise body temperature.

a Shade each of the boxes next to these terms a different colour. Use these colours to highlight or underline parts of the temperature control system in the text above.

stimulus ☐   receptor ☐   processing centre ☐   effector ☐   responses ☐

b Explain how these responses help to raise or lower body temperature.

i Producing more sweat _____

_____

ii Shivering _____

_____

① Antagonistic effectors

Blood vessels carrying blood to the skin have muscles in their walls.
These muscles can be contracted (vasoconstriction) or relaxed (vasodilation).
This is a good example of control by antagonistic effectors.

a Use the words from the box to complete the definition below.

effector	oppose	sensitive

One _____ works to _____ another. This gives more _____ control.

b The diagrams show vasoconstriction and vasodilation.
Explain how these responses help to control body temperature.

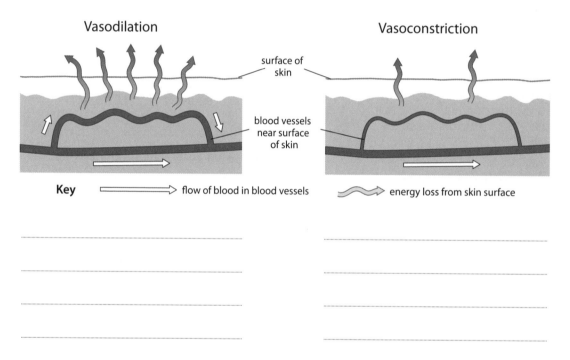

Controlling blood sugar levels

① Sugar for energy

Your cells need sugar to release energy so they can function. But too much or too little sugar can cause problems.

a The sentences below describe what happens when you have not eaten any food for several hours. Number them in the correct order. One has been done for you.

1	The blood sugar level drops.
☐	The sugar level rises quickly in your blood.
☐	You eat a packet of sweets and a chocolate bar and have a fizzy drink.
☐	You feel full of energy and restless and the hormone insulin is released into the blood.
☐	Insulin removes sugar from the blood and the blood sugar level drops to normal levels.

b Name three foods that contain starch – a complex carbohydrate.

1 .... 2 .... 3 ....

c What are the benefits of eating starchy foods compared with sugary foods?

....

d Some people do not make enough insulin or their cells do not respond to it. They have diabetes. There are two types of diabetes, Type 1 and Type 2. Look at these statements and decide whether each is true of Type 1 diabetes, Type 2 diabetes, or both. Write your answers in the last column of the table.

	Type(s) of diabetes
Develops when you are young.	
Pancreas suddenly stops making enough insulin.	
Pancreas gradually makes less insulin or cells cannot use it properly.	
Sufferers have to monitor their sugar intake.	
Normally develops in middle-to-old age.	
Controlled by insulin injections.	
Poor diet, inactive lifestyle, and obesity are risk factors.	
Symptoms include thirst and frequent urinating.	
Sufferers should have a sweet sugary snack to hand when they exercise.	
Controlled by diet and exercise.	

① Keeping fit

Exercise helps you maintain a healthy body mass.

a The table below shows the energy taken in as food and the energy used in a day, for two people.

	Jim	Kirsty
Energy intake (food) (kJ)	2100	2050
Energy used (kJ)	1754	2100
Energy gain/loss (kJ)		

i Calculate the net gain or loss in energy for Jim and Kirsty. Show an energy gain as a positive value, and an energy loss as a negative value.

ii Who will gain weight if this is a typical day? ....

iii Explain why. ....

iv Whose weight will remain roughly constant? ....

v Explain why. ....

vi If Jim always eats about the same, and always uses up the amount of energy shown in the table, what will happen over time?

....

b As well as helping to maintain a healthy body mass, suggest four other benefits of regular exercise.

1 ....

2 ....

3 ....

4 ....

c Read the text about a study of exercise in childhood and developing heart disease factors. Then answer the questions below.

Robert McMurray and his team at the University of North Carolina studied the effect of levels of fitness and physical activity during childhood on heart disease factors.

The heart disease factors were: obesity, diabetes, high blood pressure, and cholesterol.

The original group comprised 2200 children aged 7–10 years.

Seven years later the group of children had 400 participants. They were aged 14–17 years.

The scientists gathered *parallel data* on

- blood pressure
- blood fat content
- BMI
- physical activity
- aerobic fitness

Results after seven years for teenagers who had low levels of fitness and activity as children

1 Half the participants had developed one heart disease factor.

2 4.6% had developed at least three heart disease factors.

3 Of this group the children were six times more likely to have poor aerobic fitness.

4 They were five times more likely to have low physical activity.

Adapted from Early Signs Warn of Metabolic Syndrome in Teen Couch Potatoes
By Michael Smith, North American Correspondent, MedPage Today
Published: 4 April 2008

i Comment on the size of the sample when the study started.

ii Comment on the size of the study after seven years.

iii What do you think 'parallel data' means?

iv In Result 2, how many children does 4.6% represent?

v What conclusions can you make from the study? Use the words *correlation* and *cause* in your answer.

vi Suggest three questions about children developing heart disease that the study does *not* answer.

1

2

3

① Links between diet and health

In the UK most people can eat whatever they like. Most of us have more than enough to eat. In developing countries there is less choice about what to eat and some people are starving.

a Diet is a risk factor in many illnesses. Make a list of some of these illnesses.

- ....
- ....
- ....

- ....
- ....
- ....

b There is a higher incidence of diseases like colon (large bowel) cancer in the UK than in developing countries.

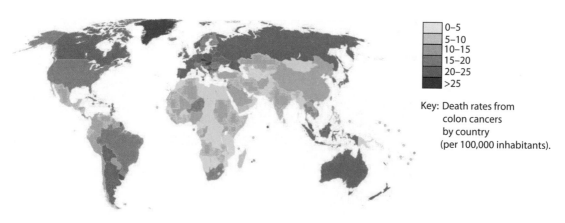

	0–5
	5–10
	10–15
	15–20
	20–25
	>25

Key: Death rates from colon cancers by country (per 100,000 inhabitants).

(Data from *Death and DAILY estimates for 2004 by cause for WHO Member States* (*Persons, all ages*) (2009-11-12); combined by Lokal Profil)

Look at the map showing deaths by colon cancer across the world.

i What conclusions can be made from the data?

....

....

ii Suggest why Europe has higher rates of colon cancer than central Africa.

....

....

iii Describe the differences between the lifestyle risk factors of people who live in Europe with its higher incidence of colon cancer and central Africa with its low incidence. Use words from the box to help you.

alcohol	cars	exercise	fruit	processed food
red meat	stress		vegetables	vegetarian

....

....

....

....

....

....

iv Suggest one way that you could show that a certain diet *caused* colon cancer?

....

....

v What are the ethical implications of what you have proposed in part **iv**?

....

Topic 4: Learning from nature

A | Linear and closed-loop systems

① Unsustainable linear systems

a Most human societies **take** natural resources, use them to **make** products, and then **dump** the waste.

   i Explain why this 'take–make–dump' process is **a linear system.**

   ....

   ii Explain why this system is not sustainable.

   ....

   ....

b Use the words in the box to complete the definition of the term 'sustainable'.

today	environment	future	needs

A sustainable system meets the .... of people ....

without damaging the .... for .... generations.

c Complete the table to show examples of what is taken, made, and dumped in a modern manufacturing process. Show what problems each stage can cause for an environment or ecosystem.

	Take	Make	Dump
What			
Problems			

② Natural closed-loop systems

a Explain why an ecosystem is often a closed-loop system.

....

b Read the passage below and then write down:

- the materials that are manufactured

- the waste products

- the reusable nutrients

> Parts of plants, such as leaves, become waste when they die. Dead leaves are eaten by snails. Faeces from snails are broken down by microorganisms, bacteria, and fungi. Microorganisms release nitrogen and phosphorus from the faeces. Nitrogen and phosphorus are taken up again by plants.

How ecosystems work

① Waste not, want not

Ecosystems are a type of natural closed-loop system; most waste is reused.
Read the information about the ecosystem badgers live in and then answer
the questions below.

> ## Badgers
>
> European badgers live in wooded areas. They live together in extensive systems of burrows called setts.
>
> Badgers live on a diet of field mice, earthworms, insects, grains, seeds, and berries.
>
> Field mice eat nuts, grains, and seeds.
>
> Earthworms digest dead plant material.
>
> When the badger dies, other animals and birds may feed on it. Insects lay eggs on the flesh and insect
> larvae eat the flesh. Microorganisms and bacteria decompose the body further. The products of
> decomposition return to the soil. These nutrients are taken up by plants.

a i What waste does the badger produce when it is alive?

....

  ii Explain what happens to this waste.

....

b Draw a diagram to show the badger in its closed-loop system.

② Closed-loop societies

The Inuit are the native people of the Arctic. A century ago, they lived in a closed-loop society. Now they are making a rapid transition into a modern society, while retaining some aspects of their traditional way of life. The table below compares aspects of the Inuit's lifestyle 100 years ago to the present day.

	Inuit in 1900s	Waste	Inuit in 2011	Waste
Home	nomadic: igloos in winter and tents made from animal skins in the summer		generally live in modern dwellings, with electricity	
Clothes	animal skins		modern clothes, some fur coats	
Transport	sledges made from bones and animal skin, pulled by dogs		cars and motorised sledges	
Food	mostly meat and fish from hunting		meat from hunting, processed food, and imported fruit and vegetables	
Death	body was wrapped in animal skins or wool blanket and placed face up in the tundra; they built cairns from stones to prevent animals eating the body, but bones were often found scattered across the tundra		cremation	

a Underline in **red** those aspects that are part of a closed-loop society and in **blue** those aspects that are a linear lifestyle.

b For both the 1900s and 2011 columns, fill in the type of waste produced by each aspect of their lifestyle and how the waste was/is reused.

c What are the advantages of modern living for the Inuit?

....

....

d What are the disadvantages for the environment of Inuit modern living?

....

....

....

Waste disposal in natural systems

1) Movement of chemicals in ecosystems

Dead organic matter (DOM) is any material that was part of a living organism. It is recycled within an ecosystem.

a In the table below give three more examples of DOM. For each one, suggest an organism that would break it down and how. An example is done for you.

DOM type	Organism that breaks it down	How
tree branch	fungi	digestive enzymes

b Carbon is a reactant that is recycled in ecosystems. Draw a flow chart showing how carbon moves between the atmosphere and plants in this cycle. Include the words in the box.

plants atmosphere glucose DOM carbon dioxide glucose carbohydrates photosynthesis cellulose bacteria starch

c Waterlogged soils have low oxygen content. Explain why plant growth is poor on waterlogged soils.

① Natural insurance policy

In nature the most important thing for a living organism is to reproduce successfully.

The male gamete usually has to travel to the female gamete (egg).

a How many sperm are needed to fertilise an egg? ....

b Suggest what happens to the sperm that are not used.

....

c Why does a bull make so many sperm?

....

② Inputs and outputs

Many ecosystems have inputs and outputs. They are not perfect closed loops.

Read the following paragraph about an ecosystem and then answer the questions below.

A lake has a stream feeding into it and another stream leaves it. One year, there was a drought and the stream feeding the lake dried up. The level of water dropped until the stream leaving the lake dried up too. People and animals that lived above and below the lake started to take water from it because there was no water in the streams. The weather also stayed very hot so water evaporated from the lake. The water level in the lake dropped quickly and fish started to die.

a At what point did the lake stop being a stable ecosystem?

....

b Explain why the lake stopped being a stable ecosystem, using the terms 'gains' and 'losses'.

....

....

c Apart from water, what other gains might the stream have brought to the lake?

....

d What effects would the lake drying up have on particular animals and plants in the ecosystem surrounding the lake?

....

....

Ecosystem services

① Natural ecosystem services

Below is a list of services that a forest provides. Draw a straight line from each service to its correct explanation.

Service	Explanation
oxygen	Provides food for human hunters.
water	Allows sexual reproduction to take place to produce variation in the next generation.
soil	Essential chemicals are recycled through the environment.
mineral nutrients	Broken down rock and decayed dead animal and plant matter holds water and nutrients for plants.
pollination	A solvent that dissolves nutrients so plant roots can absorb them.
fish and game	A gas released by photosynthesis that animals can breathe.

② Direct drilling

Ploughing damages earthworms. Direct drilling is a different method of sowing seeds/planting crops.

Explain how direct drilling, which damages fewer earthworms, benefits the soil.

....

....

① Bioaccumulation

Some rubbish produced by humans cannot be broken down. These substances are non-biodegradable.

a In the list below <u>underline</u> the waste that is *biodegradable* in **green** and the waste that is *non-biodegradable* in **blue**.

- cardboard
- plastic bags
- grass cuttings
- computer
- glass
- tea bags

- hollow-fibre pillow
- nylon carpet
- plastic bath
- wooden garden seat
- stainless steel sink
- CD-Rom disc

b Some of our waste contains chemicals that accumulate in ecosystems.
Read the passage about mercury in the sea and then answer the questions below.
The table shows the levels of mercury in ppm at each stage.

When mercury gets into the sea it is converted to methyl mercury. In this form mercury is absorbed into organisms six times faster than mercury the element.

Phytoplankton absorb the methyl mercury. The phytoplankton are eaten by zoo plankton and plant-eating fish. The zoo plankton and plant-eating fish are eaten by larger fish, birds, and humans. Humans may also eat the large fish. Bioaccumulation occurs when an organism consumes a toxin faster than it can get rid of it. Predator fish, such as tuna, swordfish, and sharks, eat a lot of prey and digest it quickly. The mercury builds up in their fatty tissue faster than their bodies can get rid of it.

If water has one unit of mercury, this table shows the amounts of methyl mercury that organisms contain as it goes up the food chain. Mercury is toxic. It affects the brain, spinal cord, kidneys, liver, and immune system. It is also absorbed by developing fetuses. Humans test foods for mercury levels so as to avoid its poisonous effects.

water		phytoplankton		zooplankton and plant-eating fish		tuna, swordfish, and sharks
1		100 000		1 000 000		10 000 000

Arbitrary units of mercury concentration in a food chain.

i Write down the definition of bioaccumulation.

....

....

ii Older, bigger fish contain more methyl mercury. Explain why.

....

....

iii At which stage is the biggest increase in methyl mercury concentration?

....

iv Explain why the threat of mercury is greater for sharks than it is for humans.

....

....

Sustainable food production

① Closed-loop farming

a In the list below, draw a ring around the words that are examples of biomass.

grass   crops   wood   aluminium cans   fish   fishing nets   humans   glass

b Here is a simple diagram of how modern agriculture often works.

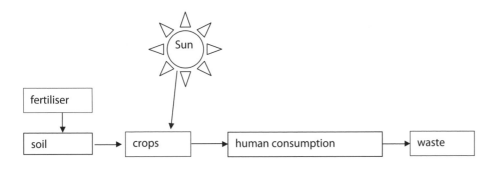

  i What sort of system is shown in the diagram?

....

  ii Draw an arrow on the diagram to show how agriculture could be made into a closed-loop system.

iii Explain what has to happen to make agriculture into a closed-loop system.

....

  iv Why might this be difficult to achieve?

....

....

② Eutrophication

Eutrophication happens when nutrients build up in the water.
The river Wonnom is well known for its fly-fishing and wildlife. In the spring of 2010 the water went green. Shortly afterwards dead fish were washed up on the banks and were floating on the surface.

a Explain what you think was happening.

....

....

....

b The environmental officer investigated the river further to find the cause of the eutrophication.

She took water samples at various points A–F along the river and drew a map. The map and a table of her results are shown below.

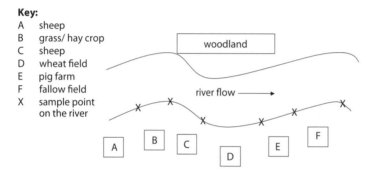

Key:
A sheep
B grass/ hay crop
C sheep
D wheat field
E pig farm
F fallow field
X sample point on the river

	sample	A	B	C	D	E	F
	1	0.20	0.19	0.16	0.82	1.60	1.38
	2	0.16	0.17	0.17	0.85	1.53	1.41
Nitrogen (mg/l)	3	0.18	0.18	0.21	0.78	1.03	1.42
	Mean						

i Work out the mean readings for each point and write them in the table.

ii Why did the environmental officer take three samples at each point?

....

iii Explain what you think has happened to the river at point E, and how you could reduce this problem through closed-loop thinking.

....

....

iv At point E one of the readings is much lower than the others. What is the name of such a reading and what would you do about it?

....

③ Deforestation – a possible solution

In 2010 Haiti was hit by a force 7 earthquake. It has also had a series of tropical storms and hurricanes. These natural disasters were made worse by unsustainable timber harvesting, clearing forests for agriculture, building, and grazing of livestock. Heavy rain combined with weak soil conditions caused mud slides and flooding.

Only 3% of forest remains. Haitians are also cutting down even more trees for firewood and charcoal because of increasing fuel costs.

a How has deforestation made the situation worse in Haiti?

....

....

b One suggestion to help Haitians is to grow a biofuel crop called *Jatropha curcus*.

It is high value, requires minimum agricultural input, grows on poor land, and reduces soil erosion.

i How could planting this new crop help Haiti?

....

....

ii What possible risk to the environment is there if a new plant species is added into the ecosystem in Haiti?

....

....

iii *Jatropha curcus* is a crop. Write a paragraph for a local magazine that explains why farmers in Haiti should avoid intensive agriculture if they decide to plant this crop.

....

....

....

....

① **Impacts of agricultural crops**

The Sahel is an area south of the Sahara desert. It has a long dry season and a shorter rainy season. In recent years the population has been using more intensive linear-system farming methods. This change from traditional methods has increased the threat of desertification.

a   Complete these sentences.

When land turns to desert this process is called ....

A definition of the term 'desert' is ....

b   The diagram shows the factors that may contribute to desertification and those that help maintain dry land where crops can be grown sustainably.

    i   Underline the factors that contribute to desertification in **red** and those that prevent it in **green**.

   ii   In each box, explain how the factor either contributes to or prevents desertification and/or flooding.

  iii   Put a **blue** border around each factor that contributes to flooding during the rainy season.

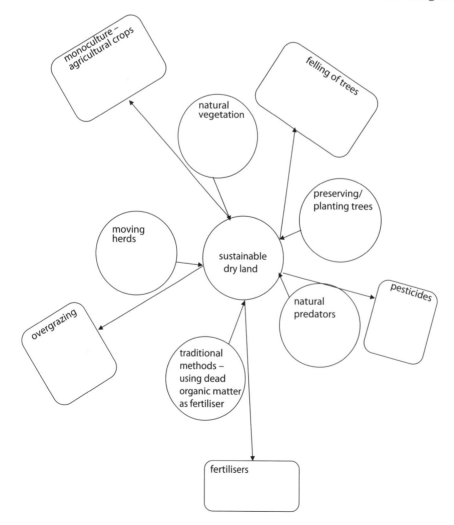

How valuable are tigers?

Biodiversity

a Write a definition of biodiversity.

....

b Conserving ecosystems protects ecosystem services for people. Look back at Section 4E and list the services that ecosystems provide.

- ....

- ....

- ....

- ....

c The tiger is under threat. The population in 2010 was only 1500; in the 1800s it was 45 000.
Calculate the percentage decline in the tiger population over this time period.

Percentage decline = ...

d Here is a food chain of a tiger: plants ⟶ samba deer ⟶ tiger

Explain the advantages and disadvantages to the ecosystem and to people of living tigers.

Advantages: ....

....

Disadvantages: ....

....

e China banned farmed tiger products, which has increased the tiger poaching in India. Why do you think the government made this ban?

....

f India recognises the value of its tigers and would like to increase their population. Suggest three measures the Indian government should take.

1 ....

2 ....

3 ....

① **Replacing what we take**

a <u>Underline</u> in **green** the correct definition of the term 'sustainable' in this list.

- Sustainable means not taking anything away.

- Sustainable means what is taken away can be quickly renewed.

- Sustainable means taking something away, and then not replacing it by the same amount or not replacing it quickly enough.

b We need to conserve forests so that people have a sustainable source of wood.

Complete the sentences below to explain how forests can be managed sustainably. Use the words in the box. You will need to use one word twice.

biomass   nutrients   trees

When timber is removed .... are cut down in the a forest.

The biomass contains ....

Sustainable use of timber means that .... and ....
have to be replaced.

c This spider diagram shows ways to manage forests sustainably. Add notes to the diagram to explain each point.

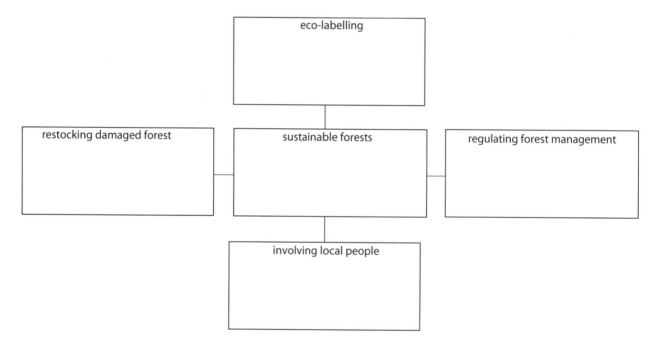

Oil, energy, and the future

① Unsustainable energy

a Oil is a fossil fuel made millions of years ago. The sentences below describe how oil formed. Number them in the correct order. One has been done for you.

☐ The oil was trapped underground.

☐ They were slowly covered up by layers of sand and silt.

1 Tiny plants in the sea made biomass from carbon dioxide and water using energy from the sun by photosynthesis.

☐ The dead bodies of the very small plants and animals fell to the sea bed.

☐ Sand and silt became rock.

☐ When the rock is drilled, the oil flows up to the surface.

☐ Some tiny plants were eaten by very small animals.

☐ Heat and pressure changed the dead biomass into oil.

b To explain more about oil, fill in the blanks in these sentences using the words in the box.

carbon	energy	hydrocarbon	made	oil
quickly	sunlight	sustainable	years	

Oil is a _____. It is a chemical made of hydrogen and _____.

The _____ stored in oil is from _____ that reached the

Earth millions of _____ ago. It takes millions of years to make _____.

Oil is being used up very _____, far faster than it is being _____.

The use of fossil fuels is not _____.

c 'An economy built on oil (as in developed countries) is a linear system.'

Under the headings below, list the evidence that supports this statement.

Take	Make	Dump
oil from the ground		

d The 'dump' part of the system is causing particular problems. Explain the problems created by dumping each of the following.

   i CO_2 ....

....

   ii Plastics ....

....

L   The Sun and sustainable living

① Sustainable energy from the Sun

a We use energy from oil to produce much of our fuel in the developed world. Think about all the ways in which oil is used in each stage of food production and add these to the diagram below. The first one is done for you.

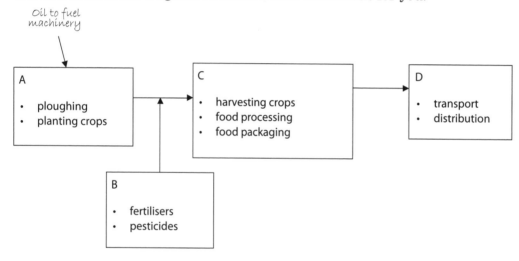

b Many parts of the world still use traditional methods without oil. Add arrows to the diagram below to show how energy flows through the system. One is done for you.

```
Sun


crops/ biomass


animals for farming          faeces

food for people
```

c Name two disadvantages of traditional methods.

   1 ....

   2 ....

② Modern closed-loop systems

a Biofuels could be an answer to closing the loop. They are sustainable fuels. What are the advantages and disadvantages of biofuel? Answer in the table below.

Advantages	Disadvantages

b Nature's closed-loop systems are sustainable. The right-hand side of the diagram below shows how they cycle biomass. We could make the way we use resources and process waste more like a natural system. If we can close the loop, perhaps our future technological society can still be sustainable. Complete the left-hand side of the diagram to show what we need to do with products to match nature's system.

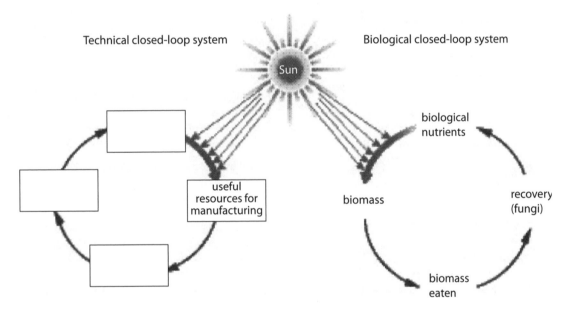

c For the 'technical closed-loop system' to work, what properties do the 'useful resources for manufacturing' need to have?

_____

d Suggest some systems that might help people to reclaim useful resources from used products.

_____

_____

e To change the 'take–make–dump' system to a 'take–reuse–recycle' system will require a complete change of culture.

Suggest how the culture needs to change, and who might be involved.

_____

_____

Topic 5: New technologies

A Living factories

① Useful chemicals from microorganisms

Microorganisms are grown in huge tanks called fermenters. A wide range of products can be made in this way.

a Here is a list of products. Decide whether the product is an *enzyme*, another kind of *protein*, or an *antibiotic*. Then draw lines to match each product to its use.

Product	Type
Quorn	

Product	Type
protease	

Product	Type
chymosin	

Product	Type
lignocellulase	

Product	Type
penicillin	

Use

breaks down lignin and cellulose in waste plant material; the sugar produced can be used to make biofuel

food

treating disease caused by bacterial infections

used in 'bio' detergents to break down protein stains

vegetarian 'rennet' used to make cheese

b List five conditions that need to be controlled in the fermenter.

- ....
- ....

- ....
- ....

- ....

c Bacteria have many features that make them ideal for industrial and genetic processes.

   i The following are steps in the process for making molecules from bacteria. Number the steps in the correct order.

ii Draw a straight line from each step to its correct feature(s).

No.	Step		Feature

| the conditions are monitored and controlled | so that it makes the correct type of molecule |

| the correct type of bacteria is carefully selected | no need for ethical concerns for bacteria |

| the bacteria are placed in the fermenter. Very soon there are large nunbers of them | vectors can carry new genes into the plasmid |

| the required molecule is extracted and the bacteria are disposed of | to produce the greatest quantity of product as possible |

| if needed, new genes are introduced into the bacteria | rapid reproduction |

Genetic modification

① New genes for cells

Genetically modified (GM) organisms are used to make new products and improve efficiency.

a In the diagram showing the genetic modification of a bacterium, add notes to describe the three main steps in the process.

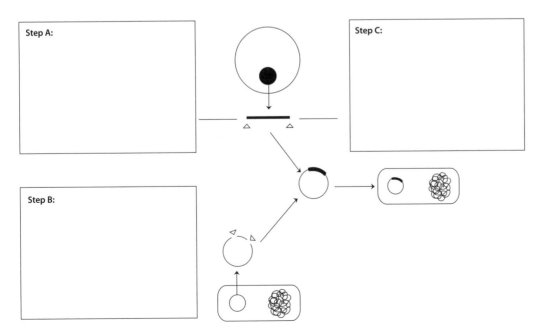

Step A:

Step B:

Step C:

b Use these words to complete the sentences.

bacteriophage	plasmid	vectors

The name for a structure that is used to insert DNA into another organism is

called a .... When DNA is inserted into a bacteria a virus

called a .... is sometimes used. The DNA is inserted into

a .... inside the bacteria.

c Outline the main steps necessary to transfer gene X from a bacterium to sugar beet using a virus vector. The gene X makes the sugar beet more disease-resistant.

• **Step A** ....

....

• **Step B** ....

....

• **Step C** ....

....

d Use two different colours to highlight the types of organism used in fermentations (box 1). Then use the colours to match them up with the useful products of fermentations (box 2). (At least one product is made two alternative ways, so could be coloured with two different colours.)

1

GM fungi or bacteria
natural fungi or bacteria

2

penicillin
insulin
alcohol
Quorn
rennin
human growth hormone

Designing for life

① Benefit and risk

Here is a summary of Philippe Vain's research on nematodes at the John Innes Centre together with the University of Leeds.

- Nematodes (microscopic worms) live in the soil and attack roots of crops. A large infestation would devastate a crop.
- Nematodes are controlled with chemicals called nematicides. These are expensive and very toxic to humans and the environment.
- Some crops have a gene for cystatin. This chemical is usually found in seeds. Insects cannot eat plant parts that contain cystatin because it affects their digestion.
- If cystatin were produced in the roots, the crops studied would be safe from nematode attack.
- Philippe Vain and his team introduced another copy of the gene responsible for making cystatin into the crop plant, using GM technology.
- The cystatin gene was expressed in the root cells and made cystatin.
- The resulting plants are highly resistant to nematodes.

Listed in the table are the main arguments against GM technology.

- For each argument decide if it applies in this case or not and explain why.
- For the arguments against GM that apply to this case, suggest a counter argument to justify the work of Philippe and his team.

Arguments against introducing genes into plants	Applies in this case? (Yes or No)	Explanation (and counter argument if you have answered 'Yes')
Added genes could make 'safe' plants produce toxins or allergens.		
Marker genes for antibiotic resistance could be taken up by disease organisms.		
Pesticides could 'leak' out of the roots of GM plants and damage insects or microorganisms that they were not designed to kill.		
GM crops may cause changes to ecosystems that cannot be reversed.		
It will cost farmers more to buy seeds of GM crops, so food costs will increase.		
Multinational organisations will increase their domination of world markets.		
Many consumers in EU countries refuse to buy GM products so farmers may lose markets.		
Poor farmers will not be able to afford the GM seeds.		

D   Genetically modified crops

① Implications of using GM organisms

There are economic, social, and ethical implications concerning the release of GM organisms.

a   The potential risks of introducing GM organisms into the environment need to be balanced against the potential benefits. Use the table to summarise some of the arguments about the examples given.

	Arguments for	Arguments against
Question 1: Should a company be allowed to make a new antibiotic against tuberculosis in a fermentation system using GM bacteria?		
economic		
ethical		
social/ environmental		
Question 2: Should farmers be allowed to grow insect-resistant maize from GM seeds?		
economic		
ethical		
social/ environmental		

b   Read about some of the implications of growing insect-resistant maize. Then answer the questions in the boxes on the next page.

A variety of maize has been genetically modified to be resistant to the European corn borer. This insect pest bores into maize plants causing them to fall over. In infested regions it can destroy 20% of the crop.

The pest is traditionally controlled using insecticide sprays. These work only during the first three days in the corn borer's life cycle, before it bores into the plant stem. The new variety of GM maize produces Bt toxin, which kills the corn borer. The Bt gene comes from a bacterium that is used as a biological insecticide by organic farmers.

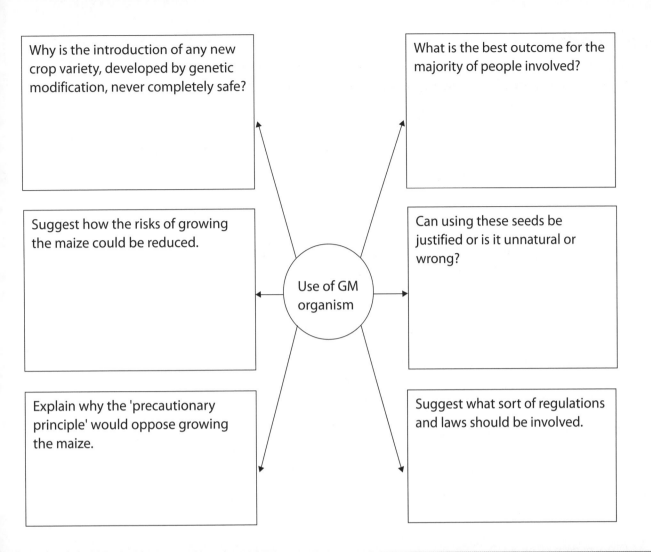

Why is the introduction of any new crop variety, developed by genetic modification, never completely safe?

What is the best outcome for the majority of people involved?

Suggest how the risks of growing the maize could be reduced.

Can using these seeds be justified or is it unnatural or wrong?

Use of GM organism

Explain why the 'precautionary principle' would oppose growing the maize.

Suggest what sort of regulations and laws should be involved.

DNA fingerprinting and profiling

1 Extracting DNA from cells

DNA for a genetic test is extracted from cells.

a Draw a simple diagram of a white blood cell, and label the features listed.

- nucleus

- cytoplasm

- cell membrane

- nuclear membrane

b **i** Where is the DNA in the cell? ....

ii Which membranes must be opened to extract the DNA? ....

iii Suggest what needs to be removed from the cell extract to leave purified DNA.

....

iv Why are red blood cells not used for DNA testing?

....

② Gene probes

a A gene probe is a short piece of single-stranded DNA that has complementary bases to a DNA chain of the test allele. Add the correct bases to make a probe for the length of DNA shown.

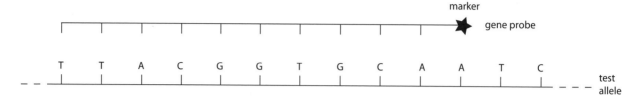

marker

gene probe

T   T   A   C   G   G   T   G   C   A   A   T   C

test allele

b Use these words to complete the description of how a gene probe works.

bind	complementary	DNA	marker

Genetic tests that use probes rely on the fact that when ....

is gently heated and cooled the double strands separate and then rejoin. If this

is done in the presence of a gene probe, the probe will ....

to sequences of DNA in the test sample that are .... to the probe.

The .... on the probe shows if this has happened.

c Markers for gene probes can be fluorescent or radioactive. Explain how the probes are detected in each case.

i Fluorescent probes: ....

....

ii Radioactive probes: ....

....

d Explain the meaning of each of these terms used in DNA technology.

- Gene ....

- Allele ....

- Gene probe ....

- UV ....

3 Developing the technique

a The sentences below describe the DNA fingerprinting technique developed by Sir Alec Jeffreys in the 1980s. Fill in the missing words using the words in the box.

band	break open	closely	complementary	dark	DNA
electrophoresis	enzymes	radioactive	unique	X-ray	

• .... the cells in the sample.

• Isolate the .... from other molecules in the cell.

• Cut the DNA into fragments using ....

• Make a probe that is .... to selected sections of the DNA.

• The probes stick to selected multiple repeats of DNA (mini-satellite sequences).

• The probes that are used are made ....

• Fragments of DNA are separated, using ....
  (gel on a plate with an electrical field across it).

• Fragments of the same size move together, and form a ....

• Bands are shown as .... lines on .... film.

• Each person has a .... number and pattern of bands.

• .... related people have some bands in common.

b Since the 1980s this technique has been developed further. Below are developments for some of the stages. Write a description of each development and explain why it is an advantage. You may need to do some research to answer this question.

Development	Description of development	Why it is an advantage
size of sample needed		
gel electrophoresis		
number of repeated sequences highlighted		
markers used		
measurements		

F   Genetic testing

① Using information

Genetic tests have been devised to provide information about a person's DNA.

a List three ways that information from genetic tests is used.

1 ....

2 ....

3 ....

b Some people would like everyone's DNA profile to be on a DNA database. What are some advantages and disadvantages of this?

Advantages	Disadvantages

G   Introducing nanotechnology

① Using tiny particles

a A nanometre is very small. Draw a ring around the number that shows how many nanometres there are in a metre.

1×10^{12} 1 000 000   1 000 000 000 1×10^{3} 1×10^{15} 1×10^{5} 10 000

b In the following list, draw a ring around what you would most likely use nanometres to measure.

A cell membranes

B skin cells

C human hair

D bacteria

c Nanoparticles have different properties to those of small particles of the same element, for example, silver. It is thought that this is because of the way that the surface-area-to-volume ratios change as particles become smaller.

Calculate and compare the surface area of these cubes.

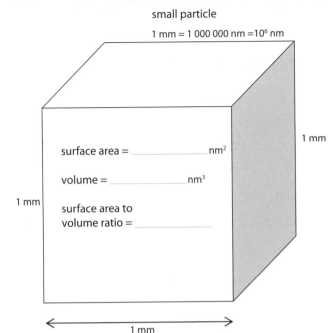

small particle

1 mm = 1 000 000 nm =10^6 nm

1 mm

1 mm

surface area = _____ nm²

volume = _____ nm³

surface area to
volume ratio = _____

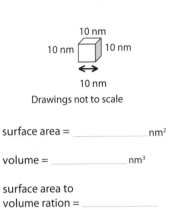

nanoparticle

10 nm

10 nm   10 nm

10 nm

Drawings not to scale

surface area = _____ nm²

volume = _____ nm³

surface area to
volume ration = _____

d Silver has antibacterial properties. Use the words in the box to explain how silver nanoparticles are used in food packaging and containers.

contamination	containers	embedded	food	fresh	wrap

_____

_____

_____

_____

_____

e Some people think that using silver nanoparticles in food packaging could be dangerous. As the food packaging breaks down, it might release the silver particles into soil or water, which could poison soil microorganisms or animals such as fish. Explain how you would decide if the risks are worth taking. Include a summary of the risks and benefits, and the decision you come to.

_____

_____

_____

_____

① Stem cells for healing

Most body cells are differentiated. They can only make copies of themselves.

a Write a definition of stem cells.

....

b There is still much to find out about the potential of stem cells. But stem cells in specialised tissues can develop into a range of cells.

Name two examples of tissues that contain stem cells.

1 ....

2 ....

c What sort of cells does bone marrow make?

....

d Scientists can grow stem cells from a patient's skin cells to make tissue to use as a skin graft.

Leukaemia sufferers have bone marrow that makes too many white blood cells. Treating leukaemia involves killing the patient's own bone marrow cells with radiation. New bone marrow introduced from a donor can make healthy blood.

How is the treatment of leukaemia different from having a skin graft?

....

....

e The table below shows some potential sources of stem cells. For each, state the disadvantages of using this source of cells.

Source	Disadvantages
umbilical cord	
embryos	
bone marrow from a donor	

Biomedical engineering

1 Mending the heart

You will need to do some research to be able to answer this question.

The heart is an amazing organ but sometimes it does not work as well as it should.

a Fill in the table below, showing what can go wrong with the heart and what symptom each problem leads to.

Problem	Symptom
	irregular heartbeat
coronary artery blocked	
	blood does not flow through heart or around body properly, causing tiredness

b The heart's pacemaker is a bundle of cells that keep the heart beating in a regular rhythm.
How does the pacemaker do this?

....

c Heart valves can become torn or stiff and may need replacing. Replacement valves can come from a donor (animal or person) or be artificial.
Give an advantage of an artificial valve.

....

d A biomedical engineer is designing a new type of heart valve. What properties must the new valve have?

....

....

....

....

....

Topic 1: Green chemistry

A The work of the chemical industry

① Chemical plants

Use the words in the box to complete the labelling of the diagram below, which summarises key aspects of a typical chemical process in industry.

| by-products | catalyst | energy | feedstocks | products |
| raw materials | reactor | recycling | separation | wastes | water |

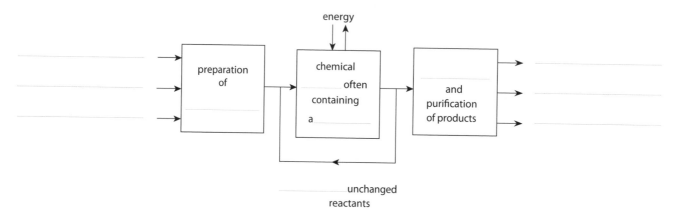

② Bulk and fine chemicals

a Decide whether each of these substances are bulk chemicals, fine chemicals, or raw materials and complete the table.

phosphoric acid　　limestone　　sodium hydroxide　　glyphosate (herbicide)　　sulfuric acid
salt　　crude oil　　citral (perfume)　　iron ore　　air　　carotene (food colouring)
ammonium sulfate (fertiliser)　　water　　ibuprofen

Bulk chemicals	Fine chemicals	Raw materials

Innovations in green chemistry

Regulation of the chemical industry

The UK government regulates chemical companies. For each of these areas, suggest a reason why it is important to have rules to protect the public, people at work, or the environment.

* Choice of raw materials for manufacturing

* Transporting chemicals

* Storing chemicals

* Getting rid of chemical waste

* Labelling of packs of chemicals

Types of feedstock

Chemical feedstocks may be either renewable or non-renewable. The grid below gives some examples. Use two colours for the key and then shade the areas in the grid with the colours to show which of these chemical feedstocks are renewable and which are not.

Key: ☐ Renewable   ☐ Not renewable

ethene from refining crude oil fractions	methanol made from natural gas and steam	sulfur from the purification of natural gas
succinic acid from the fermentation of wastes from papermaking	sodium chloride (brine) from the dissolving of underground salt deposits	ethanol from the fermentation of sugars
oxygen from the air	lactic acid from the fermentation of beet sugar	naphtha from distilling crude oil

(3) Atom economy

There are two processes for making the chemical ethylene oxide. Complete the tables for the two processes and calculate the atom economies. Some of the boxes have already been completed for you.

(Relative atomic masses: H = 1, C = 12, O = 16, Cl = 35.5, Ca = 40)

a Method 1: The multi-step route

The overall equation for method 1 is:

$$C_2H_4 + Cl_2 + Ca(OH)_2 \longrightarrow CH_2\overset{O}{\triangle}CH_2 + CaCl_2 + 2H_2O$$

Formulae of chemicals used (reactants)	Symbols of atoms in the reactants	Relative mass of atoms in the reactants	Symbols of atoms that end up in the product	Relative mass of atoms that end up in the product
Totals				

Atom economy _____

b Method 2: One-step route with a catalyst

$$2C_2H_4 + O_2 \longrightarrow 2CH_2\overset{O}{\triangle}CH_2$$

Formulae of chemicals used (reactants)	Symbols of atoms in the reactants	Relative mass of atoms in the reactants	Symbols of atoms that end up in the product	Relative mass of atoms that end up in the product
Total				

Atom economy _____

Catalysts

a Industrial catalysts speed up reactions. Why does this mean that a more efficient catalyst can allow chemical plants to operate with smaller reactors?

....

....

b Good catalysts are highly selective. Why can this help to make a chemical process 'greener'?

....

....

....

....

(5) Sustainable chemistry

The sustainability of a chemical process depends on many different factors.
Some of these factors are given below.

Draw a line from each aspect of sustainable chemistry to the example
that best illustrates it.

Aspect of sustainable chemistry   **Example**

| Avoiding chemicals that are hazardous to health |

| PET plastic waste can be depolymerised, turning it back into the monomers originally used to make the polymer. The result is fresh feedstock for making new polymer. |

| Managing energy inputs and outputs |

| A weedkiller manufacturer used to use hydrogen cyanide as a starting material. Hydrogen cyanide is toxic. They now use a new method with a less harmful starting material. |

| Recycling chemicals |

| The process for making the pigment titanium dioxide also makes iron sulfate and gypsum. Iron sulfate can be used to treat water and gypsum can be used to make plasterboard. |

| Finding new uses for by-products |

| The original method for making ibuprofen had a low atom economy. A newer process has a much higher atom economy. |

| Converting more of the reactants to products products |

| A chemical company used an infrared survey to find areas where heat was being lost from steam pipes so that they could insulate them more effectively. |

Topic 2: The chemistry of carbon compounds

The alkanes

Methane molecules

Complete this table, which shows three ways of representing a molecule of methane. Add these terms in the correct places as column headings:

- Ball-and-stick model
- Molecular formula
- Structural formula

CH_4		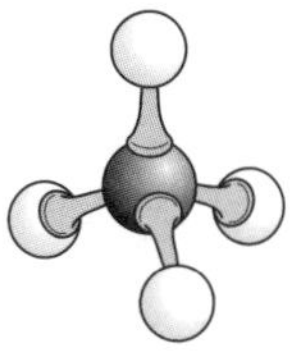

Alkane formulae

Complete this table by adding the missing information about three alkanes.

Name	Molecular formula	Structural formula
	C_2H_6	
		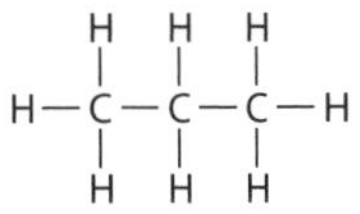
butane		

③ Burning alkanes

Many fuels contain alkanes. Alkanes burn in air. They react with oxygen. Write a balanced equation, with state symbols, for the reaction of ethane burning in plenty of air.

④ Alkane properties

Use the terms in the box to complete the summary of the properties of alkanes. Use each word only once.

| alkane | crude oil | gases | hydrocarbons | oily | single bonds | water |

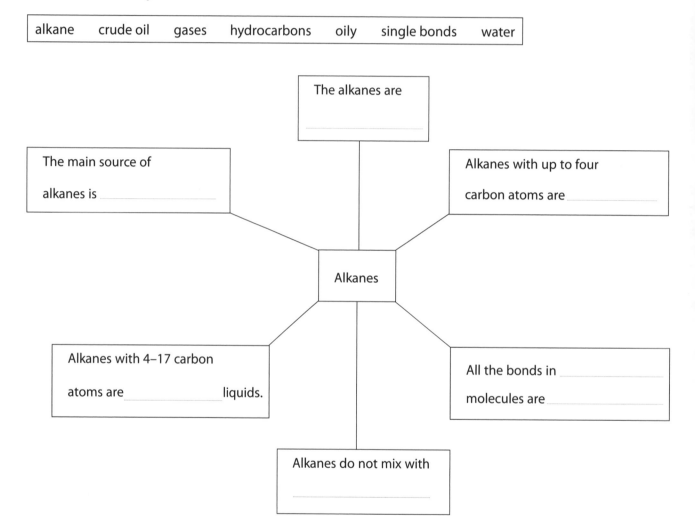

The alkanes are _____

The main source of

alkanes is _____

Alkanes with up to four

carbon atoms are _____

Alkanes

Alkanes with 4–17 carbon

atoms are _____ liquids.

All the bonds in _____

molecules are _____

Alkanes do not mix with

_____

Alkane structure and reactivity

Add notes and labels to this diagram to explain why alkanes do not react with common laboratory reagents, such as acids and alkalis.

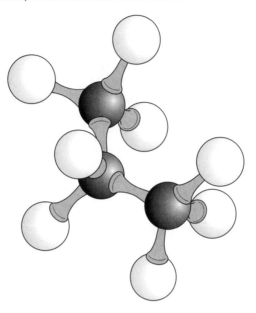

① Methanol and ethanol

Complete this table. In the models, colour the carbon atoms black and the oxygen atoms red. Leave the hydrogen atoms white.

	Methanol	Ethanol
Molecular formula		
Ball-and-stick model		
Uses		

Alcohols compared to water and alkanes

Complete this table to compare methanol with water and methane. In the models, colour the carbon atoms black and the oxygen atoms red. Leave the hydrogen atoms white.

	Water	Methanol	Methane
Molecular formula			
Ball-and-stick model			
State at room temperature			
Boiling point		65°C	−161°C
Ease of mixing or dissolving in water			
Explanation of ease of mixing/ dissolving in terms of the strength of the attraction between molecules			

③ Functional groups

In this diagram of an ethanol molecule, colour the carbon atoms black and the oxygen atoms red. Leave the hydrogen atoms white. Then label and annotate the diagram to show:

- the functional group
- the bonds that are reactive
- the bonds that are not reactive

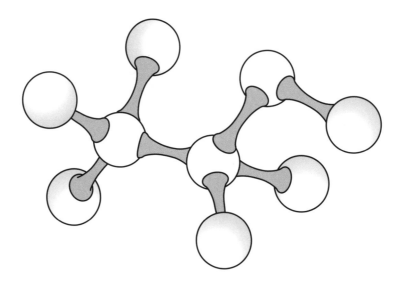

Reactions of alcohols

a In some ways alcohols are like water. However, alcohols burn and water does not. Explain why alcohols can burn and state the products of burning if there is plenty of air.

b Write a balanced equation for the reaction of ethanol burning in air.

c Sodium is a reactive metal.

   i In the second column of the table, put a tick for each of the chemicals that react with sodium.

Chemical	Does the chemical react with sodium?	Names of products	Formulae of products
water			
methanol			
methane			

   ii For each of the chemicals that react with sodium, complete the row by writing in the names and formulae of the products.

The production of ethanol

① Uses of ethanol

Give examples to illustrate these uses of ethanol made on an industrial scale.

a As a fuel ....

b As a solvent ....

c As a feedstock for other processes ....

② Making ethanol by fermentation of sugars

Feedstocks

a Give one example of a purpose-grown crop and one example of a waste material that might each be used as a raw material to make ethanol from biomass.

    1 A purpose-grown crop ....

    2 Waste material ....

Fermenting sugars with yeast

Fermentation converts sugars into ethanol and carbon dioxide. Enzymes in yeasts catalyse the reactions. Yeast is a living organism. Fermentation is an example of anaerobic respiration.

b Write a balanced equation for the fermentation of glucose ($C_6H_{12}O_6$).

....

c Fermentation with yeast works best at temperatures in the range 25–37°C. Suggest reasons why fermentation is slow:

    i below this temperature range ....

    ii above this temperature range ....

d Why does fermentation slow down or stop if the alcohol concentration exceeds 14%?

....

e How is it possible to obtain ethanol solutions with a concentration above 14%?

....

Making ethanol from biomass with bacteria

Feedstocks

Breaking down biomass with acid can produce a wide range of sugars. These include six-carbon sugars such as glucose ($C_6H_{12}O_6$), and five-carbon sugars such as xylose ($C_5H_{10}O_5$). Yeast is good at converting six-carbon sugars to ethanol but not five-carbon sugars.

a Give two reasons why there is a need to find new ways to convert sugars to alcohol.

   1 An economic reason ....

   ....

   2 An environmental reason ....

   ....

Fermenting sugars with GM bacteria

Scientists have used genetic modification to create a bacterium that can convert five-carbon sugars to ethanol. The diagram below shows a process that makes ethanol from biomass using this GM bacterium.

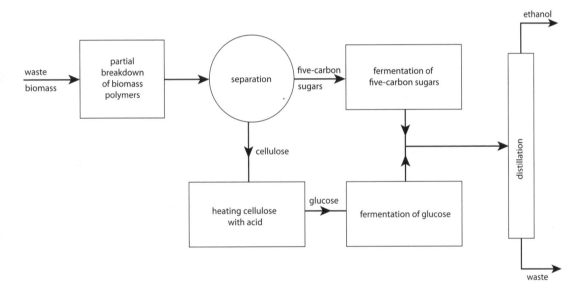

b What is the purpose of heating the cellulose with acid? ....

   ....

c Suggest a reason for fermenting glucose in a separate tank from xylose and

   other five-carbon sugars. ....

   ....

d What is the purpose of the distillation? ....

   ....

4 Making ethanol from petrochemicals

One method of making ethanol uses ethane, a petrochemical. The sources of ethane are natural gas and the distillation of crude oil.

Producing the feedstock, ethene

a Heating ethane at a high temperature breaks up the molecules and produces ethene, C_2H_4, and hydrogen. Write a balanced equation for the reaction.

....

Making ethanol from ethene

A mixture of ethene and steam under pressure combines to make ethanol in the presence of a phosphoric acid catalyst. About 5% of the mixture is converted to ethanol as the compressed gases pass through the catalyst.

b Complete the equation for this reaction.

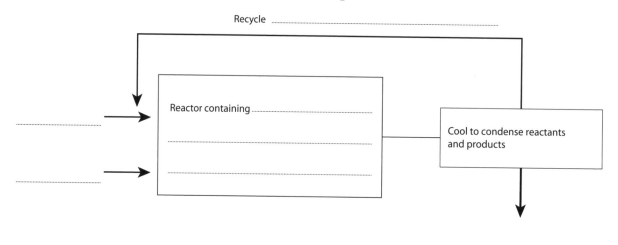

c What is the theoretical atom economy for this reaction? ....

d Complete the labelling of this flow diagram for the process.

Recycle ....

Reactor containing ....

....

Cool to condense reactants and products

e Why is it necessary to recycle ethene in the process?

....

f Why does the ethanol produced need to be purified?

....

g Overall, the yield of ethanol from ethene is 95% of the theoretical yield. What mass of ethanol would be produced from 50 tonnes of ethene?

Ethanol manufacture and the environment

The diagram below compares, in outline, fuel ethanol from crops with fuel ethanol from crude oil. Fuel ethanol can be described as carbon neutral if the carbon dioxide given out when the fuel burns is balanced by the amount absorbed at stages during the production of the fuel. Some crop-based ethanol can be carbon neutral.

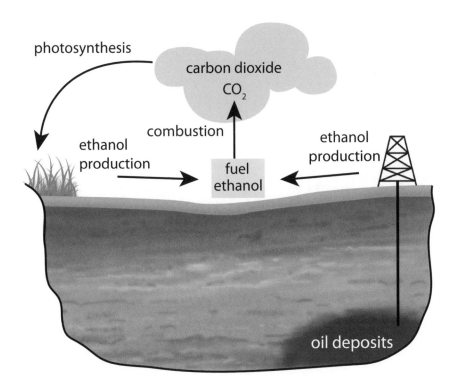

a Why is fuel ethanol made from crude oil not carbon neutral?

b Why is it desirable to use fuels that are carbon neutral? ....

....

c Suggest one reason why crop-based ethanol cannot be completely carbon neutral in practice.

....

....

d Identify two disadvantages of growing crops to make crop-based ethanol.

1 ....

2 ....

Carboxylic acids

1 pH of solutions of acids

Draw lines to match each solution to its pH value. Some solutions listed have the same pH value.

Dilute acetic acid (ethanoic acid)		pH 7
Dilute hydrochloric acid		pH 3
Vinegar		pH 1
Pure water		

2 Reactions of carboxylic acids

Complete these word equations to show the typical reactions of organic acids with metals, metal hydroxides, and metal carbonates.

methanoic acid + ... → magnesium methanoate + hydrogen

methanoic acid + sodium hydroxide → .... + ...

ethanoic acid + ... → copper ethanoate + water

ethanoic acid + potassium carbonate → .... + ... + ...

③ Carboxylic acid formulae and structures

a Complete this table. In the models, colour the carbon atoms black and the oxygen atoms red. Leave the hydrogen atoms white.

	Methanoic acid	Ethanoic acid
Molecular formula		
Structural formula		
Ball-and-stick model		

b What is the functional group in an organic acid?

_____

c **i** Write a symbol equation to show how ethanoic acid ionises when it dissolves in water.

_____

ii Which of the products of the ionisation of ethanoic acid makes the solution acidic?

_____

d Draw a ring around the formulae below that are carboxylic acids.

CH_3OCH_3 HCOOH $CH_3CH_2CH_2CH_2OH$ $CH_3CH_2CH_2COOH$

Esters

1 Esters

a Give one word to describe the typical smell of many simple esters. ....

b Give examples of three foods we eat that taste of mixtures of esters.

1 .... 2 .... 3 ....

c Give two uses of esters.

1 ....

2 ....

d Use the names of the chemicals in the box below to write a word equation for the formation of an ester.

| water   ethanol   ethyl butanoate   butanoic acid |

....

e Complete this word equation.

pentanol   +   ethanoic acid   →   ....   + ...

f Label the diagrams below to show how to make a small sample of an ester and then smell the product.

The words and phrases in the box will help you.

| acid   alcohol   catalyst   carboxylic acid   hot water
neutralise leftover acid   reaction mixture after warming   sodium carbonate solution |

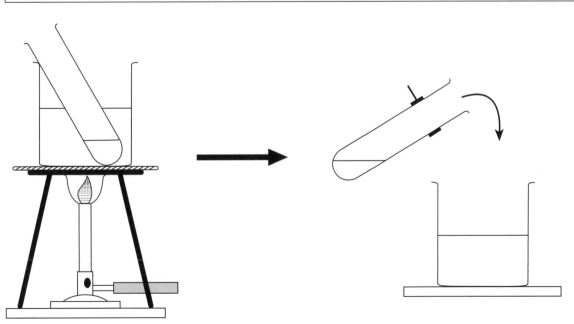

② Preparation of an ester

This flow diagram shows the procedure used to make a pure sample of an ester on a laboratory scale.

Annotate (label) the diagram. The words in the box below will help you. You may choose to use some words more than once.

aqueous reagent to remove impurities   catalyst   drying agent   ethanoic acid   ethanol ester from tap funnel   heat   impure ester layer   concentrated sulfuric acid   distillation conical flask   impure product   pure ethyl ethanoate   reflux condenser   tap funnel thermometer   condenser   shake tap funnel

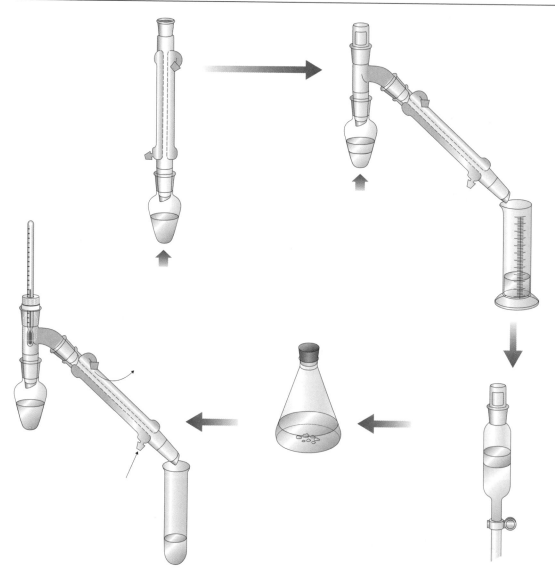

Fats and oils

① Fats and oils

a Why are fats and oils important to plants and animals?

....

b These diagrams show the structures of a molecule from a fat and a molecule from an oil. Label both diagrams to show: an ester link, a hydrocarbon chain, and the part of the molecule that comes from glycerol.

Molecule A

Molecule B

c Are the hydrocarbon chains in molecule A saturated or unsaturated? Is this the structure of a molecule from a fat or an oil? Give your reasons.

....

....

....

....

d Are the hydrocarbon chains in molecule B saturated or unsaturated? Is this the structure of a molecule from a fat or an oil? Give your reasons.

....

....

....

....

① Exothermic and endothermic reactions

A student carried out an experiment. She mixed sodium hydroxide solution with hydrochloric acid and found that the temperature of the reaction mixture increased.

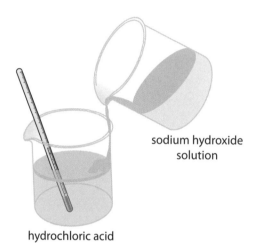

sodium hydroxide solution

hydrochloric acid

a She decided to mix together other chemicals, as shown in the table below, to see if there was also a temperature change. She measured the temperature of the reaction mixture at the start and the end. Complete the following table to show the change in temperature and whether the reactions were endothermic or exothermic.

	Chemicals mixed	Temp at start (°C)	Temp at end (°C)	Change in temp (°C)	Endothermic or exothermic
A	Hydrochloric acid and zinc	19	25		
B	Citric acid, sodium hydrogencarbonate, and water	18	10		
C	Water and potassium chloride	19	17		
D	Magnesium and lead nitrate solution	18	35		

b Complete the sentences below by putting a ring around the correct bold word.

In an **exothermic/endothermic** reaction the temperature rises. The reaction **gives/takes** energy **in from/out to** the surroundings.

In an **exothermic/endothermic** reaction the temperature decreases. The reaction **gives/takes** energy **in from/out to** the surroundings.

2 Exothermic reactions

a The reaction of magnesium with dilute hydrochloric acid is an exothermic reaction. Use the words in the box to label the diagram below, and explain what is meant by the term 'exothermic reaction'.

dilute hydrochloric acid   energy given out   exothermic
magnesium   magnesium chloride solution

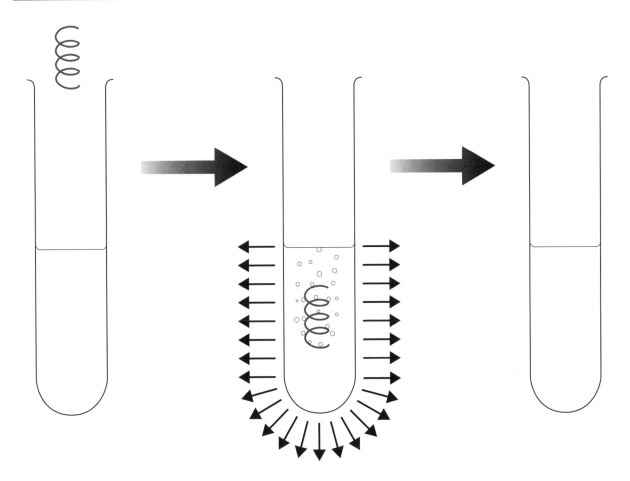

b Write a balanced symbol equation, with state symbols, for the reaction of magnesium with dilute hydrochloric acid.

....

c Use parts of the symbol equation to label the energy-level diagram below by adding the reactants and products to the diagram. Complete the labelling of the diagram.

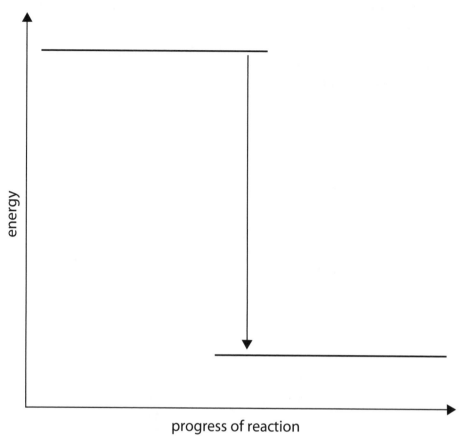

progress of reaction

d Give two more examples of exothermic changes.

1 ....

2 ....

Endothermic reactions

a The reaction of citric acid with sodium hydrogencarbonate is an endothermic reaction. Use the words in the box to label the diagram below, and explain what is meant by the term 'endothermic reaction'.

citric acid   endothermic   energy taken in   sodium citrate solution
sodium hydrogencarbonate

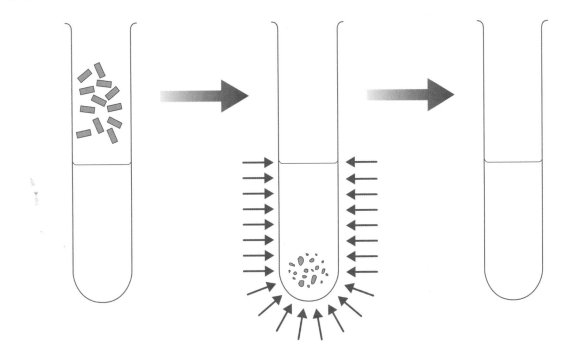

b Write a word equation, with state symbols, for the reaction of citric acid with sodium hydrogencarbonate.

c Use parts of the word equation to label the energy-level diagram below by adding the reactants and products to the diagram. Complete the labelling of the diagram.

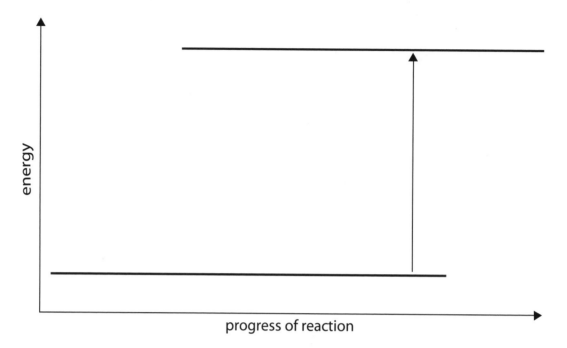

progress of reaction

d Give another example of an endothermic change.

Bond breaking and bond forming

a Use the words in the box to label the diagram below. Colour the oxygen atoms red; leave the hydrogen atoms white.

bond broken during reaction	bond formed during reaction	hydrogen molecule
oxygen molecule	water molecule	

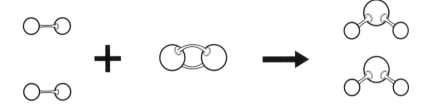

b Explain why the reaction of hydrogen with oxygen is an exothermic reaction.

....

....

c Hydrogen (H_2) reacts with bromine (Br_2) to form hydrogen bromide (HBr). Use the data in the table to complete the labelling of the diagram on the next page.

Process	Energy change for breaking all the bonds in the formula mass of the chemical
Breaking the H–H bonds in hydrogen	434 kJ needed
Breaking the Br–Br bonds in bromine	193 kJ needed
Breaking the H–Br bonds in hydrogen bromide	366 kJ needed

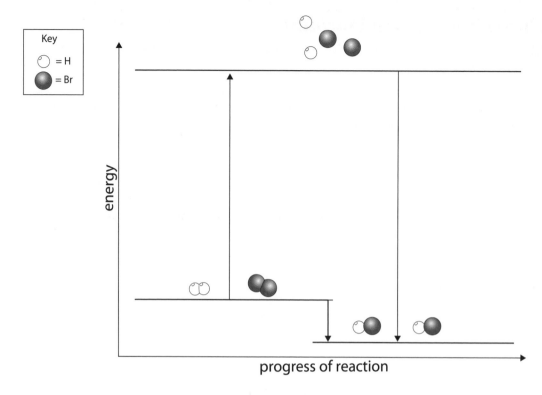

d Calculate the overall energy change for the reaction between hydrogen and bromine.

....

....

How fast?

Activation energies

a With the help of some or all of the words and phrases in the table, write a paragraph to explain what is meant by the term 'activation energy'.

new bonds form	high energy	bonds break	unsuccessful collision
reactant molecules	millions of collisions per second	low energy	collisions
not all collisions lead to reaction	minimum energy	successful collision	product molecules

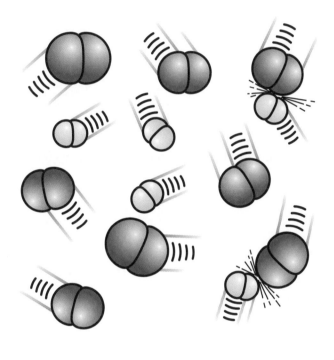

....

....

....

....

....

....

....

b Use the idea of activation energy to explain why reactions go faster as the temperature rises.

....

....

....

....

....

c Catalysts provide a different route for a reaction with a lower activation energy. Why does this make the reaction go faster?

....

....

....

....

d The diagram below shows the energy profile for an exothermic reaction.

i Label the diagram to show the activation energy for the reaction.

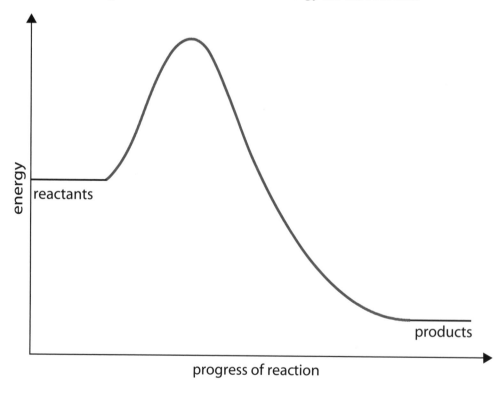

ii A catalyst provides an alternative route for a reaction. Draw and label a new energy profile on the same axes to show the effect of using a catalyst to speed up this reaction.

Topic 4: Reversible reactions and equilibria

Reversible changes

Reversible changes

a For each of these pairs of equations, indicate the conditions needed to make the change go in the direction shown in the equation.

Conditions needed to make the change happen in the direction shown:

i $H_2O(l) \rightarrow H_2O(g)$

$H_2O(g) \rightarrow H_2O(l)$

ii $CuSO_4.5H_2O(s) \rightarrow CuSO_4(s) + 5H_2O(l)$

$CuSO_4(s) + 5H_2O(l) \rightarrow CuSO_4.5H_2O(s)$

iii $NH_3(g) + HCl(g) \rightarrow NH_4Cl(s)$

$NH_4Cl(s) \rightarrow NH_3(g) + HCl(g)$

b Hot iron reacts with steam to form iron oxide, Fe_3O_4, and hydrogen.

i Write a balanced symbol equation for the reaction.

ii This reaction is reversible. Label the diagram to show how to demonstrate the reverse reaction.

iii Write a balanced symbol equation for the reverse reaction.

① **Dynamic equilibrium**

a The diagram below shows a crystal of iodine dissolving in hexane. The solution formed is then shaken with a solution of potassium iodide in water. Colour the solutions.

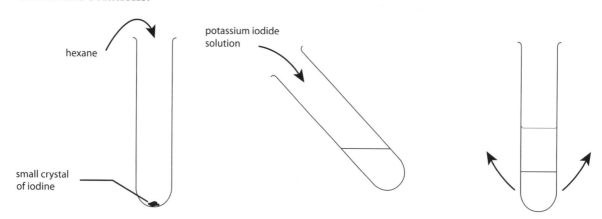

Explain why there is some iodine in both layers, however long the two solutions are shaken up together. ....

....

....

b The diagram below shows a crystal of iodine dissolving in aqueous potassium iodide. The solution formed is then shaken with hexane. Colour the solutions.

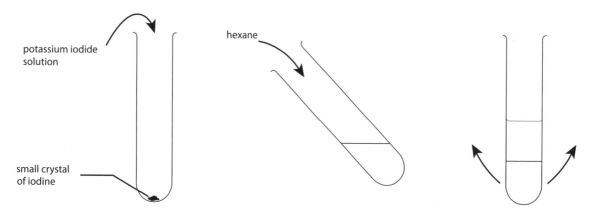

Explain why the two layers look the same after shaking as they did when the crystal was first dissolved in hexane and then shaken with potassium iodide solution.

....

....

c Why is the term 'dynamic equilibrium' used to describe the state reached when solutions of iodine in hexane and in aqueous potassium iodide are shaken up together?

....

....

Ammonia and green chemistry

Making ammonia

Ammonia is made by the Haber process. The equation for the reaction is:

$$N_2(g) + 3H_2(g) \rightleftharpoons 2NH_3(g)$$

a What does the symbol $\rightleftharpoons$ mean?

....

b How many molecules of gas are there on the left-hand side of the equation? ....

c How many molecules of gas are there on the right-hand side of the equation? ....

d The reaction of nitrogen with hydrogen is exothermic. Will the reverse reaction

(to break up the ammonia) give out heat or take in heat? ....

The graph below shows the percentage yield of ammonia at different temperatures and pressures.

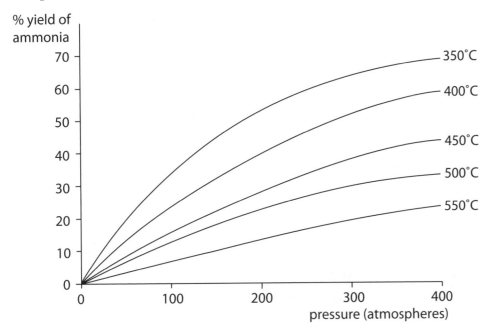

e What happens to the yield of ammonia as the pressure increases?

....

f Use ideas about reversible reactions to explain your answer to part **e**.

....

....

....

....

g What happens to the yield of ammonia as the temperature increases?

....

h Use ideas about reversible reactions to explain your answer to part **g**.

....

....

....

....

i What will happen to the rate of the reaction as the temperature increases?

....

j A temperature of 450 °C is usually chosen for the reaction. Suggest reasons why this temperature is chosen by considering what will happen if the temperature is higher or lower.

....

....

....

....

k What else is added to the reaction chamber to increase the rate of the reaction and improve the yield?

....

② Making the Haber process greener

The Haber process is a relatively clean chemical process but there are ways it could be made greener.

a Give two examples of ideas that scientists are exploring to try to make the Haber process greener. Explain how each example might help to reduce the environmental impact of making ammonia.

Example 1 ....

....

....

Example 2 ....

....

....

Topic 5: Chemical analysis

Stages in an analysis

Analysts at work

Read these descriptions of events that involved analysts.

1   A professional cyclist tested positive for clenbuterol, a banned substance. The cyclist must have taken the drug to enhance their performance. The cyclist was banned from competing for two years.

2   There was an algal bloom in a river. The water contained high levels of ammonium nitrate, which is used as a fertiliser. The algal bloom must have been caused by local farmers allowing fertiliser into the water course.

3   Sudan 1 is a dye that is used for colouring solvents, waxes, and shoe polish. It is not allowed to be added to food in the UK. A sample of chilli powder was found to be contaminated with Sudan 1 dye. Many ready meals were found to contain Sudan 1 too. The chilli powder presumably all came from the same batch and the dye was probably added to it before it reached the UK.

a   For each of the three events, identify the statements that *report data*. Underline these in red.

b   For each of the three events, identify statements that are *explanatory ideas* (hypotheses, explanations, and theories). Underline these in blue.

c   What was the role of the analyst in each event and why was their involvement important?

Event 1 ....

....

....

Event 2 ....

....

....

Event 3 ....

....

....

① Sampling for analysis

a When taking samples for analysis, it is important to make sure that they are typical of the whole bulk of the material being analysed. Analysts have to decide:

- how many samples to collect, and how much of each to collect, to be sure that the samples are representative

- how many times to repeat an analysis on a sample to be sure that the results are reliable

- where, when, and how to collect samples of the material

- how to store samples and take them to the laboratory to prevent samples 'going off', becoming contaminated, or being tampered with.

Write down the main issues for the person planning the analysis of the following materials.

i Soil in a farmer's field

ii Water in a pond

iii Polluted air in a city street

iv Urine from an athlete

b Give reasons why it is important to have standard procedures for collecting, storing, and analysing samples.

Chromatography

Paper and thin-layer chromatography

Paper and thin-layer chromatography can be used to analyse mixtures of chemicals.

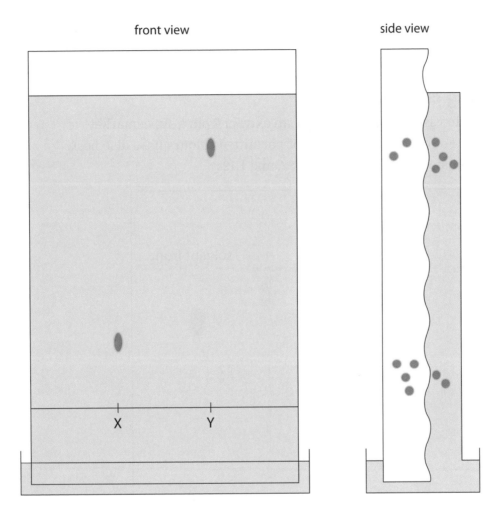

front view   side view

a Label this diagram as described in parts i – iii. It may help to use colours.

  i Label the mobile phase, stationary phase, and solvent front.

 ii Add an arrow to show the direction in which the solvent front moves.

iii Add arrows to show that for both chemicals in samples X and Y there is a dynamic equilibrium between the stationary phase and the mobile phase.

b **i** What would happen to a spot of substance on the start line that is not at all soluble in the mobile phase?

....

ii Explain why sample Y moves further than sample X.

....

....

② Interpreting chromatograms

The diagram below is a chromatogram of an extract from a supermarket curry sauce (S). Four reference samples of permitted colours have also been run on the chromatogram (E102, E110, E122, and E124).

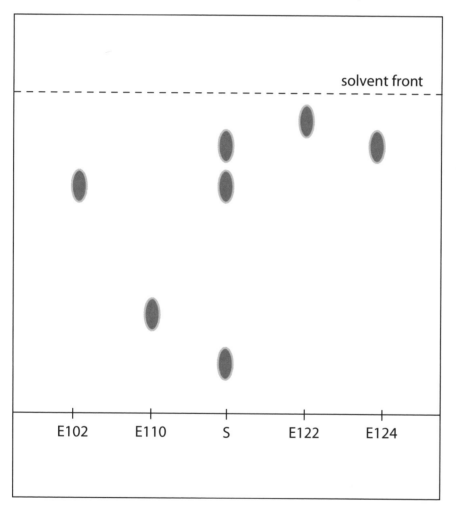

a How many coloured chemicals were there in sample S? ....

b Which permitted additives were present in the curry sauce? ....

c Calculate the R_f value for the spot that does not match any of the reference colours.

$$R_f = \frac{\text{distance moved by spot}}{\text{distance moved by solvent front}}$$

....

....

d Under the same conditions, the $R_f = 0.15$ for a banned colouring, Sudan 1. What does the chromatogram show about the colourings in the curry sauce?

....

....

e Suggest two methods that could be used to detect colourless additives on the chromatogram.

1 ....

2 ....

① Gas chromatography

a The diagram below describes gas chromatography (GC). Use the words in the box to label the diagram.

chromatogram column packed with stationary phase cylinder of carrier gas detector
flow meter mobile phase oven recorder sample injected here vent

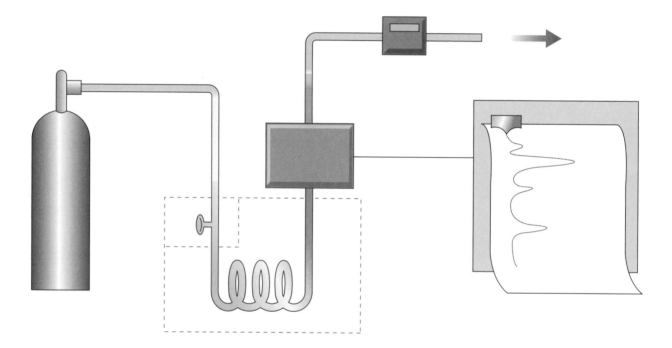

b The diagram below shows a gas chromatogram.

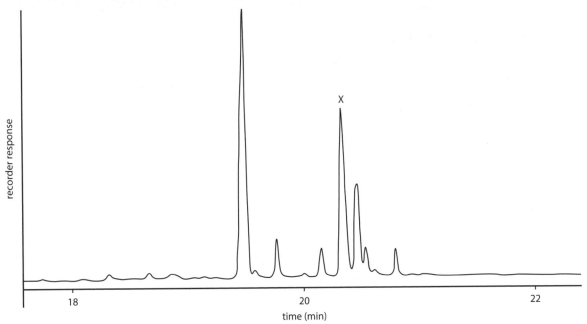

i How many chemicals were there in the mixture injected into the

GC instrument? ....

ii Label the peak that is caused by the chemical present in the largest amount
in the sample.

iii Estimate the retention time of chemical X. ....

....

① Concentrations of solutions

a Label this sequence of diagrams and complete the captions to show how you would prepare a solution of sodium carbonate, Na_2CO_3, with a concentration of 10.6 g/dm³, assuming that the volume of the graduated flask is 250 cm³.

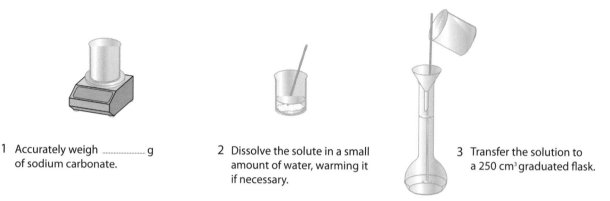

1 Accurately weigh .... g of sodium carbonate.

2 Dissolve the solute in a small amount of water, warming it if necessary.

3 Transfer the solution to a 250 cm³ graduated flask.

4 Rinse all the solution into flask with more ....

5 ....

6 ....

b Complete the table to show the concentrations of these solutions in g/dm³.

Solution	Concentration (g/dm³)
20.0 g magnesium sulfate in 500 cm³ solution	
4.5 g potassium hydroxide in 250 cm³ solution	
0.5 g sodium sulfate in 10 cm³ solution	
1.25 g silver nitrate in 50 cm³ solution	

c Complete the table to show the mass of solute in these volumes of solutions.

Sample of solution	Mass of dissolved solute in (g)
100 cm^3 of a 25.0 g/dm^3 zinc sulfate solution	
50 cm^3 of a 10.0 g/dm^3 lead nitrate solution	
10 cm^3 of a 22.5 g/dm^3 magnesium chloride solution	
2.5 cm^3 of a 16.0 g/dm^3 barium nitrate solution	

2) Use of a pipette

A pipette is only accurate if it is used correctly. The following questions are designed to remind analysts about correct checks and procedures. Suggest reasons for each question being asked.

a Have you rinsed the pipette with the solution you are going to measure out?

_____

_____

b Have you made sure that there are no air bubbles in the narrower parts

of the pipette? _____

_____

_____

c Have you wiped the outside of the pipette to remove solution on the outside

of the glass before running out the liquid? _____

_____

_____

d Have you lined up the meniscus with the graduation mark correctly?

_____

_____

③ Use of a burette

A burette is only accurate if it is used correctly. The following questions are designed to remind analysts about correct checks and procedures. Suggest reasons for each question being asked.

a Have you checked that the burette is clean before you start?

b Have you rinsed the burette with the solution you are going to measure before filling it?

c Have you read the burette correctly and taken both readings?

d Have you left a drop hanging from the tip of the burette after running the solution into the flask?

F   Evaluating results

① A titration to analyse vinegar

a The diagrams below show steps in a titration to measure the concentration of acetic acid (ethanoic acid) in vinegar. Use the words in the box to label the diagrams. You may use the words more than once.

conical flask   burette   indicator   pipette   sodium hydroxide   vinegar

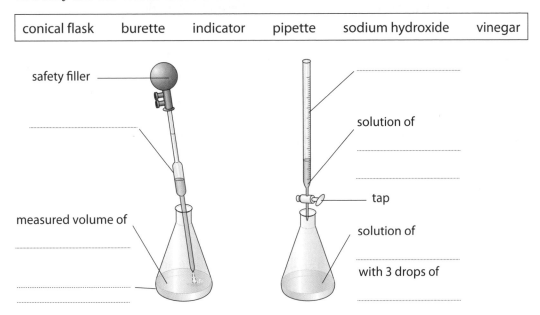

safety filler

solution of

tap

measured volume of

solution of

with 3 drops of

b The table below shows the results of a series of titrations to measure the concentration of acetic acid in vinegar. The flask contained 10.0 cm³ vinegar and three drops of phenolphthalein indicator. The concentration of the sodium hydroxide solution in the burette was 20.0 g/dm³.

	Rough titration	Titration 1	Titration 2	Titration 3
Second burette reading (cm³)	17.5	22.00	19.00	20.10
First burette reading (cm³)	0.0	5.00	2.10	3.20
Volume of NaOH(aq) added (cm³)				

i Complete the bottom row of the table.

ii Draw a ring around the values you should use to work out an average value.

iii The average value for the volume of alkali added = ... cm³

iv Use this formula to work out the concentration of acetic acid in the vinegar.

Concentration of acetic acid (g/dm³)

$$= \frac{3}{2} \times \text{NaOH concentration (g/dm}^3\text{)} \times \frac{\text{volume of NaOH (cm}^3\text{)}}{\text{volume of vinegar (cm}^3\text{)}}$$

Concentration of acetic acid in the vinegar = ...

2 Interpreting titration results

An analyst carried out a titration to find the concentration of limewater. Limewater is a saturated solution of calcium hydroxide, $Ca(OH)_2$ in water. The analyst measured out 20.0 cm^3 samples of limewater and then carried out titrations with dilute hydrochloric acid. The concentration of the acid was 1.46 g/dm^3 HCl(aq). The average titre was 25.0 cm^3 of the dilute hydrochloric acid. Follow these steps to work out the concentration of calcium hydroxide in limewater.

a Write the balanced equation for the reaction that takes place during the titration.

....

b Work out the relative formula masses of calcium hydroxide and hydrochloric acid. (Relative atomic masses: Ca = 40, O = 16, H = 1, Cl = 35.5)

$Ca(OH)_2$ ....

HCl ....

c Calculate the mass of HCl in the 25.0 cm^3 of the dilute hydrochloric acid added from the burette.

....

....

d Use the equation and the reacting masses to calculate the mass of calcium hydroxide that reacts with the HCl added from the burette.

....

....

e Your answer to part d is the mass of calcium hydroxide in 20.0 cm^3 of limewater. Calculate the concentration of calcium hydroxide in limewater in g/dm^3.

....

....

Accurate quantitative analysis

Below are stages in a quantitative analysis. For each stage, show how this applies to a titration or explain why it is important.

- Measuring out accurately a specified mass or volume of the sample

- Working with replicate samples

- Dissolving samples quantitatively

- Measuring a property of the solution quantitatively

- Calculating a value from the measurements

- Estimating the degree of uncertainty in the results

Topic 1: Naked-eye astronomy

A | What can we see in the sky?

① Sun and Moon

Both the Sun and the Moon rise in the east, and cross the sky before setting in the west.

This sequence of diagrams shows the Moon moving east to west across the sky, but slipping back though the pattern of stars.

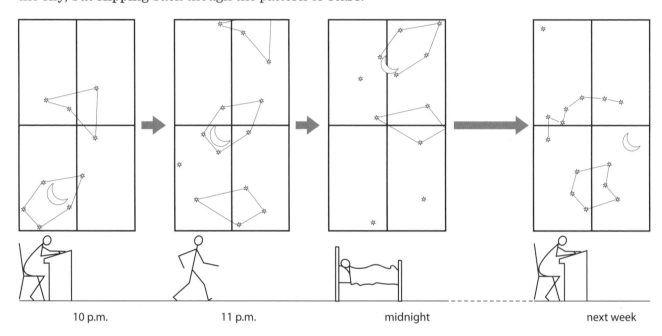

| 10 p.m. | 11 p.m. | midnight | next week |

a Fill in the time period, to complete the statement.

The Sun appears to travel east to west across the sky once every ....

b Draw a ring around the correct **bold** words to complete these statements.

  i The stars appear to travel east to west across the sky once in a slightly **longer** / **shorter** time period than the Sun.

  ii The Moon appears to travel east to west across the sky once in a slightly **longer** / **shorter** time period than the Sun.

c Explain why this happens, mentioning the Earth's rotation and the Moon's orbit of the Earth.

....

....

....

....

Phases of the Moon

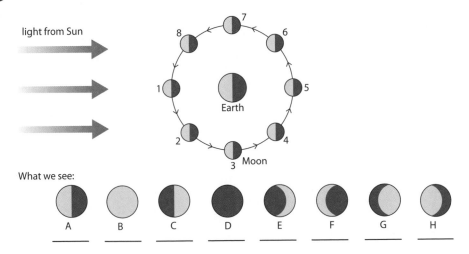

a Match diagrams A–H showing what we see with positions of the Moon 1–8 as it orbits the Earth.

b Explain why the appearance of the Moon changes in a regular way, in terms of the relative positions of the Sun, Moon, and Earth.

....

....

....

Constellations and seasons

Different star constellations are visible in the night sky at different times of the year.

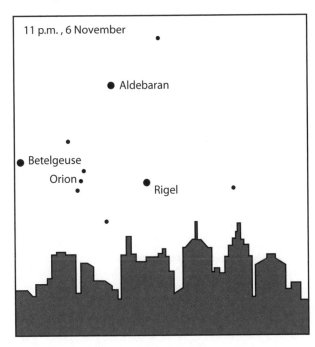

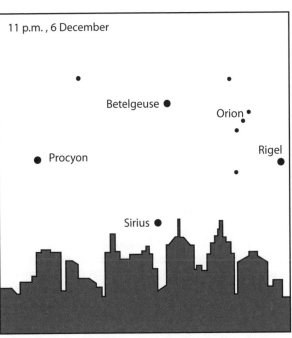

The diagram below shows the Earth orbiting the Sun and the constellations of the zodiac.

a Complete the labels on the diagram.

b Explain the observation on page 101, by referring to your labelled diagram

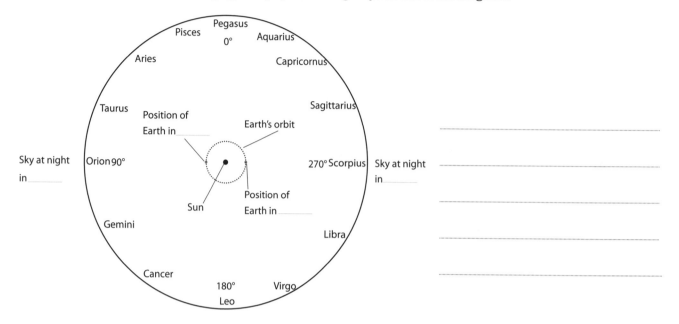

The zodiac is the set of constellations located around the ecliptic, that is, around the extended plane of Earth's orbit of the Sun.

④ **The motion of stars**

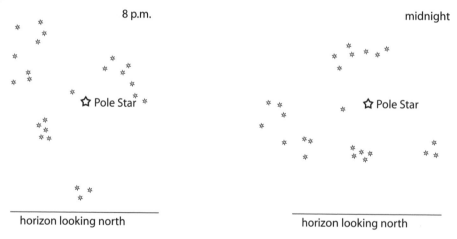

a Describe how the stars appear to rotate during this four-hour period.

....

b Explain why the stars appear to rotate.

....

....

c Why does the Pole Star not appear to move?

....

5 Sidereal and solar days

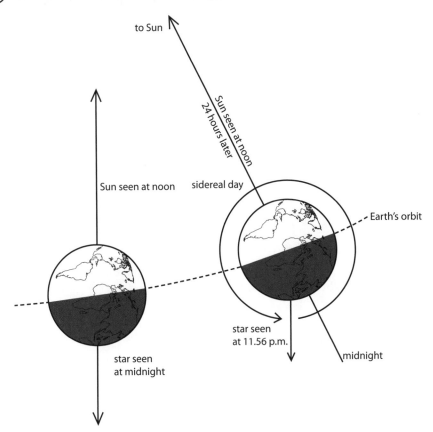

Use the diagram to explain why a sidereal day is four minutes shorter than a solar day.

....

....

6 The position of the star

Draw a diagram and explain how you could use the two angles, right ascension and declination, to describe the precise position of an astronomical object.

① Retrograde motion of planets

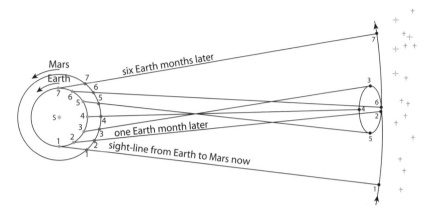

This diagram shows the position of Mars and Earth in their orbits and how Mars appears to move as seen from the Earth.

a Complete the description of this motion using words from the list. Words can be used once, more than once, or not at all.

> forwards   retrograde   backwards   sideways

From months 1 to 3, Mars appears to move forwards. Then, for two

months, it goes .... before moving .... again.

This is known as .... motion.

b Explain why Mars appears to move backwards (west to east) in the night sky for two months.

....

....

....

....

c Tick ✓ all the statements that are true.

☐ Venus and Mercury sometimes show retrograde motion.

☐ Neptune and Uranus never show retrograde motion.

☐ Mars, Jupiter, and Saturn sometimes show retrograde motion.

☐ Mercury can only be seen at dawn or dusk.

☐ Venus can only be seen in the middle of the night.

Eclipses

Eclipses

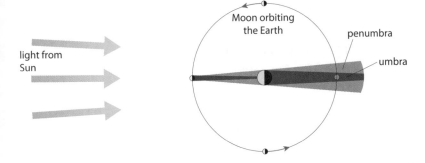

a Label the Moon's position:

X where it causes a solar eclipse

Y where it causes a lunar eclipse

b Use the diagram to explain why lunar eclipses can be seen by many more people than solar eclipses.

....

....

....

② Why eclipses are rare

This diagram shows why eclipses are not seen every month (not to scale).

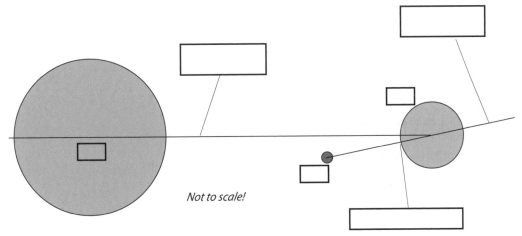

Not to scale!

a Complete the labels for the diagram.

b Use the diagram to explain why eclipses are rare.

....

....

....

A | Discovering what's 'out there'

① Observatories and telescopes

An observatory is a site used for observing objects in space. A telescope is any instrument that collects radiation from these objects for astronomers to study.

Complete the table, giving examples of observatories or telescopes and their locations.

Observatory or telescope	Type of electromagnetic radiation detected	Location
	radio and microwave	
	infrared	
	visible light	
	ultraviolet	
Chandra	X-ray	
Fermi	gamma ray	

② How telescopes work

Telescopes can make things visible that cannot be seen with the naked eye.

Describe two different ways that they might do this. Use words from the box in your description.

distant source	telescope	weak radiation
detector	part of the electromagnetic spectrum	

....

....

....

....

Making images of distant objects

1 An image from a pinhole

The diagram shows a pinhole camera.

a Draw rays from the top and bottom of the lamp to show how light travels from the lamp through the camera.

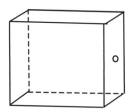

b Draw the lamp as it appears on the screen.

c Label the diagram using words from the box.

| camera | image | object | pinhole | ray | screen |

2 Describing an image

a Draw a ring around the correct **bold** words to complete these statements.

Light travels out from the **image** / **object**, but only **rays** / **lines** in the right direction pass through the pinhole. These rays form an **image** / **object** on the screen. The image is **upright** / **inverted**.

Moving the camera away from the object makes the image **larger** / **smaller**. Making the camera body longer makes the image **larger** / **smaller**.

The image on the screen is **real** / **virtual,** so it could be recorded by a light-sensitive detector or photographic film (unlike the image in a mirror, which is **real** / **virtual**.)

b Describe what would appear on the screen:

i if there were three pinholes

....

ii if there was one large hole

....

① Converging lens shapes

Describe how the shape of a stronger lens differs from the shape of a weaker lens.

....

....

② The focal length of a lens

Complete the diagram to show what happens to the rays after passing through the lens.

Label the principal axis, focus, and focal length of the lens.

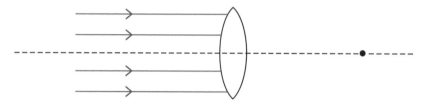

③ Power of a lens

If you know the focal length of a lens, you can calculate its power using this equation:

$$\text{lens power} = \frac{1}{\text{focal length}}$$

(.....................)

        (.....................)

a Complete the equation by filling in the units.

b Calculate the power of a lens with each of these focal lengths:

    i 0.5 m

    ii 40 cm

    iii 20 cm

    iv 5.0 m

Water waves

The diagram below shows how water waves behave when they move from deep water to shallow water.

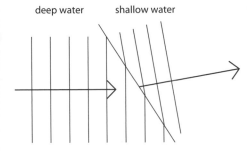

deep water   shallow water

a Draw a ring around the correct **bold** words to complete the statement below.

This effect is called **refraction / diffraction**. It can cause the waves to change **direction / frequency**. **All / Some** waves behave like this when they enter a medium in which they travel at a different **speed / frequency**. Light waves travel more slowly in **glass / air** than in **glass / air** so they are **diffracted / refracted** at air–glass and glass–air boundaries. Lenses change the direction of light rays because **reflection / refraction** occurs as light enters and leaves the lens.

b The picture below right shows a water wave going from deep water to shallower water. Choose words from the box to complete the sentences.

shallow	deep	frequency
wavelength	shorter	longer
slowed	reduced	

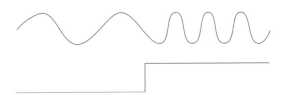

Water waves travel faster in _____ water than in _____ water.

As the wave crosses the boundary, the _____ of the wave stays the

same but the wavelength gets _____ . This is because the waves

are _____ down; the progress of each wave is _____

by the time the next wave arrives.

c The wave in the picture is refracted.

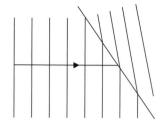

　　i Put a tick ✓ in the region where the waves have a shorter wavelength.

　　ii Are the waves slower or faster in this region? _____

　　iii Label the two regions 'slow' and 'fast' to describe the speed of the waves.

　　iv The arrow shows the direction of the wave before the boundary. Draw another arrow to show the direction of the wave after it has crossed the boundary.

　　v The statements below explain why its direction changes. They are out of sequence. Use arrows to join the boxes in the correct order. The first one has been done for you.

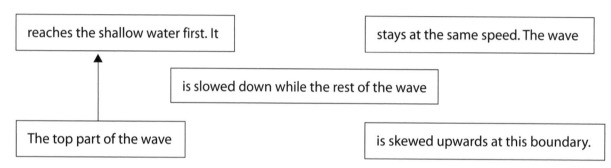

⑤ Parallel rays

Explain why light from any star arrives at a telescope as 'parallel rays'.
Use a diagram to help.

The image formed by a converging lens

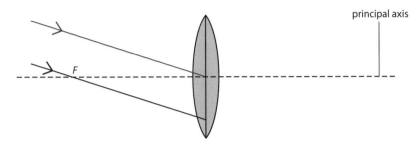

principal axis

a The diagram above shows two rays from a distant star. One strikes the centre of the lens. The other passes through the lens focus before striking the lens. Show what happens to each ray, and add labels to describe the rule for each of them.

b Label the place where the lens produces an image of the star.

c Draw two more rays coming from the same star, and passing through the lens.

d Is the image virtual or real? Explain why.

....

....

An extended object

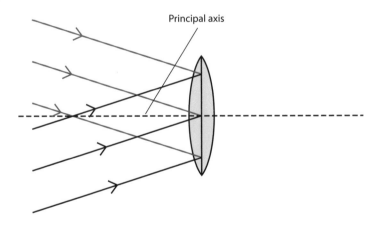

Principal axis

a The diagram above shows rays from a distant galaxy. The telescope objective lens gathers light from two sides of the galaxy. Complete the rays to show how the lens produces an image of the galaxy.

b Annotate the diagram to explain the result.

c A galaxy is one example of an extended object that an astronomer might study. Give another example.

....

① Lenses in a telescope

a Label the eyepiece and objective lens in the diagram of a telescope below.

b Describe how the shapes of the two lenses compare.

....

....

② The aperture of a telescope

The aperture of a telescope is the light-gathering area of its objective lens. Give one reason why a telescope with a larger aperture is better than one with a smaller aperture.

....

....

③ The magnification of a telescope

a Explain what is meant by the magnification of a telescope.

....

....

b Although no telescope can make stars look any larger, a telescope with greater magnification is still better for observing stars. Explain why.

....

....

c The magnification of a telescope can be calculated using this relationship:

$$\text{magnification} = \frac{\text{focal length of objective lens}}{\text{focal length of eyepiece lens}}$$

Complete the table below by calculating the magnification of a refracting telescope with each pair of lenses.

Telescope	Focal length of objective lens	Focal length of eyepiece lens	Magnification
A	50 cm	5 cm	
B	80 cm	5 cm	
C	5.0 m	5 cm	
D	1000 mm	4 mm	
E	2000 mm	10 mm	

d Which telescope provides the greatest magnification? ....

Comparing lenses

The table below lists lenses, all made from the same type of glass, that may be used in making a telescope.

Lens	Focal length (mm)	Lens diameter (mm)
P	500	80
Q	250	120
R	25	60
S	50	100

a Which lens is the most powerful? ....

b Which lens would be thinnest (have surfaces with the least curve)? ....

c Which lens, if used as the objective lens, would give the brightest image? ....

d Calculate the magnification of a telescope made using lenses P and R.

....

....

① A spectrum

The picture shows white light from a
filament lamp passing through a prism.

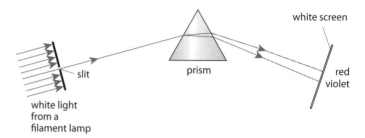

a Complete these sentences. Draw a ring around the correct **bold** words.

The light is **refracted** / **diffracted** in the prism. This produces a
spectrum / **reflection** of colours.

The speed of light in air is the same for all the colours. When it enters glass the
light slows down – it is **reflected** / **refracted**. The speed is different for different
colours in glass so the change in direction is **different** / **the same** for different colours.

b Which colour is the slowest in glass? ....

② Prisms and Gratings

a Describe two differences between the white-light spectra produced by a prism
and a diffraction grating.

1 ....

....

2 ....

....

b Draw a ring around the correct bold words to complete these statements.

Although glass is **opaque** / **transparent**, thick glass absorbs some visible light
and is **opaque** / **transparent** to ultraviolet radiation. In a spectrometer, more
of the light reaches the screen to produce a spectrum when using a **grating** / **prism**.

When astronomers are looking at spectra of starlight they need as **little** / **much**
light as possible to be transmitted by the spectrometer. This is why most astronomers
use **gratings** / **prisms** rather than **gratings** / **prisms** to look at the spectra of starlight.

Lenses or mirrors?

Reflecting telescopes

Referring to the diagram below, explain a major problem with refracting telescopes.

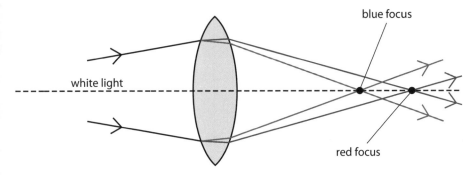

blue focus

white light

red focus

....

....

Mirrors for telescopes

Give three more reasons why most telescopes used by professional
astronomers have mirrors as their objectives, and not converging lenses.

1 ....

2 ....

3 ....

Parabolic reflectors

A parabolic mirror is the most common shape for the objective of a reflecting telescope.

Complete the diagram to show what happens to the rays after striking the mirror.
Label the principal axis, focus, and focal length of the mirror.

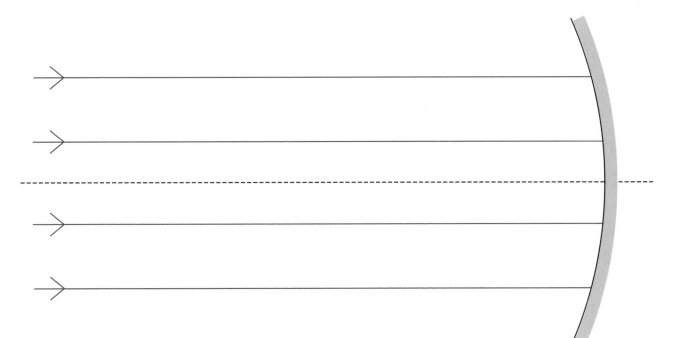

① Wave behaviour of light

Water waves can be used to model light waves.

a Water waves will spread out when they pass through a gap in a barrier.

wave approaching barrier

wave after it has passed through

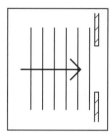

i What is the name of this effect?

....

ii The picture on the right shows some waves approaching barriers and the waves after they have passed through the barrier.

Draw lines to match each barrier with the wave after it has passed through.

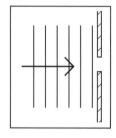

iii What happens to the spread of the wave as the gap gets wider?

....

....

....

b Complete these sentences. Draw a ring around the correct **bold** words.

Diffraction is most noticeable when the width of the slit is about the same size as the **wavelength** / **amplitude** of the wave. Light has a very **long** / **short** wavelength. A light wave will diffract when it passes through a very **narrow** / **wide** slit.

② Wavelength and diffraction

Complete the diagram below to show what happens to the waves at these apertures.

Wavelength and resolving power

Use words from the box to complete these sentences.

resolving power	diffraction	separate	electromagnetic	aperture	wavelength

The .... of a telescope is its ability to distinguish between two

closely spaced objects so that they are recognisable as .... objects.

When the telescope .... is large in relation to the .... of

the .... radiation being collected, astronomers will see more detail in an

image. This is because it will reduce the effect of wave .... at any opening.

Images and resolving power

These diagrams show two light sources, observed through three different telescopes.

Draw lines to match each image to the telescope that made it.

High-resolution telescope

Medium-resolution telescope

Low-resolution telescope

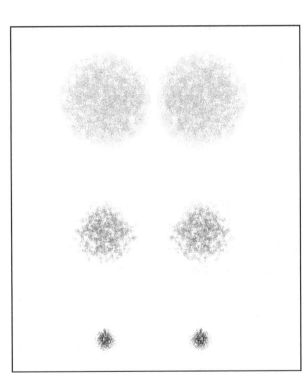

Large telescopes

Give two reasons why astronomers need to build very large telescopes.

1 ....

2 ....

Refracting telescope objective

Explain why the objective of a refracting telescope should have a large diameter but be a weak lens.

....

....

....

① Electromagnetic radiation through the atmosphere

Look at the diagram. It shows the percentage of each type of radiation that reaches the ground from outside the Earth's atmosphere.

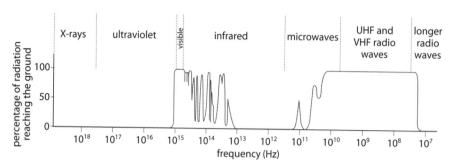

a Put an 'A' under the types of radiation that are absorbed by the atmosphere.

b Put a 'T' under the frequencies that are transmitted by the atmosphere.

c Suggest two types of scientist who make use of the microwave 'window' in the atmosphere.

  i .... **ii** ....

② Telescopes in space

Read the following statements about using telescopes outside the Earth's atmosphere. Put an A next to those that are advantages, and a D next to those that are disadvantages.

Advantage or disadvantage?	Statement
	They avoid absorption and refraction effects of the atmosphere.
	They are expensive to build and launch.
	They can detect radiation from astronomical objects in parts of the electromagnetic spectrum that are strongly absorbed by the atmosphere.
	Servicing relies on space programmes that astronomers cannot control.
	Orbit allows imaging from all parts of the sky.
	If things go wrong it is much harder to repair them.
	Instruments can quickly become out of date and are not easily replaced.
	Launching limits the size of the telescope.

Choosing a site

There are some important factors to look for when choosing the site for a major astronomical observatory.

a Draw one line from each box stating a factor to link it to the correct reason why it is important.

Factor	**Reason**
high elevation	less light pollution
frequent cloudless nights	less atmosphere to pass through
low pollution and dry air	less absorption and scatter by chemicals and water vapour
a long way from cities	less absorption and scatter by water molecules

b Draw a ring around the **four** sites on Earth where optical and infrared observatories are mostly found.

Australia   Switzerland   Canary Islands   Chile   Hawaii   Kenya   Scotland   Turkey

Twinkle, twinkle

Explain why stars appear to twinkle.

....

....

Astronomers working together at ESO

Computer control in observatories

Tick all the phrases below that complete this statement correctly.
Tick more than one.

Computer control enables a telescope to:

automatically track a distant source, collecting weak radiation from it, while the Earth rotates ☐

scan a distant source, collecting data from each part of it ☐

see clearly even when viewing conditions are poor ☐

follow instructions from an astronomer who is not based at the observatory (be operated remotely) ☐

be set to move quickly and point precisely to another part of the sky ☐

record and process data collected ☐

② Astronomy today

a Describe two ways that astronomers work with local or remote telescopes.

1

2

b Give three reasons why international collaboration is common in astronomy.

1

2

3

③ Building and operating an observatory

List four factors that are not to do with the quality of the images (non-astronomical factors) that are important in building and operating an observatory.

1

2

3

4

Topic 3: Mapping the Universe

How far away are the stars?

Parallax

Using an everyday example, explain why nearby stars appear to move during the year against the background of distant stars.

Parallax angle

The diagram below defines parallax angle.

a Label the diagram.

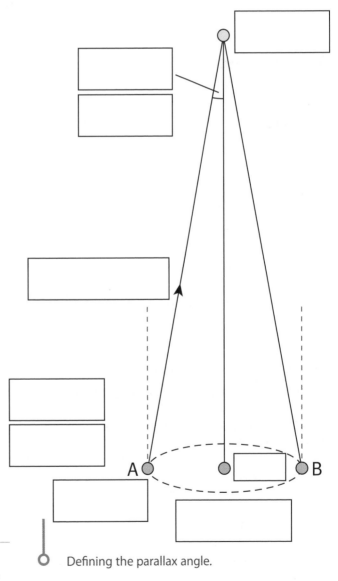

A   B

Defining the parallax angle.

b How long does it take the Earth

to move from A to B? ....

c Complete this definition of the parallax angle.

The parallax angle is ....

③ Parallax angle and distance

a Draw a diagram showing the parallax angle of two stars.

   i A star that is near to the Earth.

   ii A star that is further away from the Earth.

b Which star has the largest parallax angle?

_____

④ Star distances

Complete the following sentences.

There are _____° (degrees) in a full circle.

There are _____' (minutes) of arc in 1°.

There are _____' (seconds) of arc in 1°.

A unit of distance based on the measurement of parallax is the _____ .

An astronomical object at a distance of 1 parsec has a parallax angle of _____ .

This distance is similar to another unit (about three times larger) used for measuring astronomical distances, which is a _____ .

⑤ Parallax and parsecs

Complete the table by calculating each distance in parsecs.

Parallax angle (seconds of arc)	Distance (parsecs)
0.769	
0.1	
0.025	
0.0125	
0.06	
0.01	

Using the angle of parallax

An astronomer regularly observes a star through a telescope. The maximum angle he has to turn the telescope through after six months to see the star again is 1.2".

a Draw a (ring) around the correct **bold** values. (You may find it helps to look at the diagram in the answer for Question 2).

   i When the telescope is turned through 1.2" after six months, the value of the angle of parallax is:

0.3"	0.6"	1.2"	2.4"	4.8"

   ii This means that the distance to the star is:

0.833 parsecs	0.6 parsecs	1.67 parsecs	1.2 parsecs

b A second star has an angle of parallax of 0.8".

   i Compare the distance to the two stars.

   ....

   ....

   ii A third star does not have an angle of parallax, it does not appear to move. What does this tell you about this star, compared to the other two?

   ....

   iii Explain why angles of parallax are not used to measure distances to the most distant stars.

   ....

   ....

Brightness and distance

This diagram shows a model that a student set up to show how the brightness of a light is affected by the distance between the observer and the light source.

The identical light sources are each at the centre of a sphere.

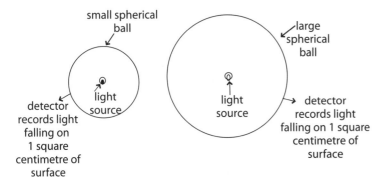

a Complete this description of how the distance from the light source affects the brightness.

As the distance from the light source increases

b Use the diagram to explain why the distance from the light source affects the

brightness.

c The student observes these facts about two street lights.

- The street lights are identical. When standing close to each light the brightness is the same.

- When looking along the street, the closer light looks brighter than the distant light.

- When smoke from a bonfire blows across the street both streetlights look dimmer than without the smoke.

i Use the example of the street lights to explain how astronomers use a star's

brightness to tell how far away it is.

ii For this method to work write down two important factors that must be the same.

1 _____   2 _____

(8) Luminosity

The rate at which a star radiates energy is called its luminosity. Underline two factors that can affect a star's luminosity:

size   twinkling   parallax angle   temperature

(9) Observed brightness of a star

Explain why the observed brightness of a star (seen by an astronomer) will depend on both its distance and its luminosity. It may help to draw a diagram.

Comparing Betelguese and Rigel

The constellation Orion contains the bright stars Betelgeuse and Rigel.

Betelgeuse

Rigel

The following statements about these stars are all true.

- Betelgeuse and Rigel are both about the same size.
- Betelgeuse is much cooler than Rigel.
- Rigel gives out more than four times as much energy every second as Betelgeuse.
- From Earth, Betelgeuse appears slightly brighter than Rigel.

Use this information to compare the distance from Earth of the two stars.

....

....

....

How far away are galaxies?

Cepheid variable stars

Cepheid variable stars are important in working out the distances to other galaxies.

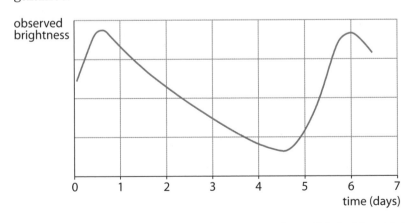

a The graph shows the brightness curve for a Cepheid variable star.
Describe the pattern in its observed brightness.

....

....

b Henrietta Leavitt made a very important discovery about Cepheids. Draw a line linking the two variables that she found were related.

luminosity
temperature
distance

observed brightness
period
speed of recession

c The following sentences describe how astronomers use a Cepheid variable star to work out the distance to another galaxy. Number the boxes to put them in the correct order. The first one has been done for you.

☐ Measure and record the brightness of the Cepheid variable star at different times.

☐ Use the period of variation of the Cepheid variable star to estimate its luminosity.

☐ Plot a graph of brightness against time for the Cepheid variable star and measure its period of variation.

☐ Knowing both the luminosity of the star and brightness, calculate the distance to the Cepheid variable star.

1 Look for a Cepheid variable star in the galaxy of interest.

☐ Take the distance of the Cepheid variable star as the distance to the galaxy.

In practice, astronomers would calculate the distance to many Cepheid variables stars in a galaxy to estimate its distance.

② The distance to a galaxy

This graph shows how the brightness of a Cepheid variable star in a distant galaxy changes over two weeks.

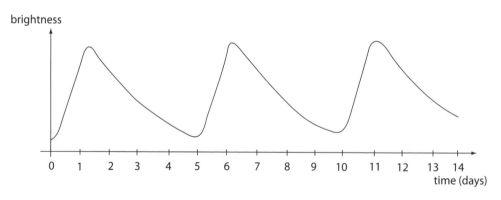

a Complete this statement.

The period of the variation in brightness is ....

b The graph below shows the luminosity of Cepheid variable stars plotted against their period in days.

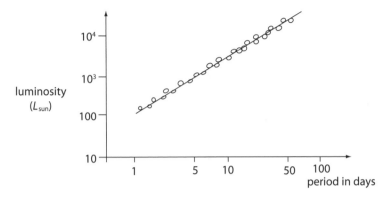

i Draw lines on the graph to show how you would use it to estimate the luminosity of the Cepheid variable star in part **a**.

ii Complete this statement.

The luminosity of the Cepheid variable star is ....

c Explain how the distance to the galaxy can now be worked out.

....

....

....

How are stars distributed in space?

The great debate – Shapley versus Curtis

a Use words from the box to complete the sentences.

nebulae	Curtis	Universe	Milky Way	galaxies	Shapley

In 1920 there was a famous debate between two American astronomers, Harlow

Shapley and Heber Curtis, about the nature and size of the ....

Central to the debate was the interpretation of thousands of fuzzy objects observed

in the night sky, which were called .... The Milky Way includes nebulae

and is much larger than previously thought, suggested ....

Spiral nebulae are outside the .... , and are distant ....

similar to our own, suggested ....

b Tick ✓ the statement that best completes this conclusion.

The debate was settled by Edwin Hubble's observations a few years later because:

☐ Shapley agreed that he had been wrong.

☐ Curtis agreed that he had been wrong.

☐ Hubble was a better scientist than Curtis or Shapley.

☐ Hubble presented new evidence based on data collected from a new telescope.

c Explain how Cepheid variable stars helped to resolve this debate. Include the name of the astronomer responsible for this breakthrough.

_____

_____

_____

d Complete these statements about the final outcome of the debate:

Curtis was correct that _____ .

The debate shows us how difficult it is to decide what the evidence

shows when _____ .

② Galaxies

The *Spitzer Space Telescope* is an orbiting infrared telescope. It was used to survey 30 million stars in the Milky Way, from which a group of astronomers were able to build up a picture of our galaxy, published in 2005.

a What is a galaxy?

_____

b Suggest why the survey was done by a *group* of astronomers.

_____

_____

c Complete the following sentences, using units in the box.

parsecs (pc)	kiloparsecs (kpc)	megaparsecs (Mpc)

Distances between stars in a galaxy are typically measured in _____ .

Distances between galaxies are typically measured in _____ .

What else is 'out there' – and where is it?

Objects in the Solar System

This is a diagram of the Solar System. It is not to scale.

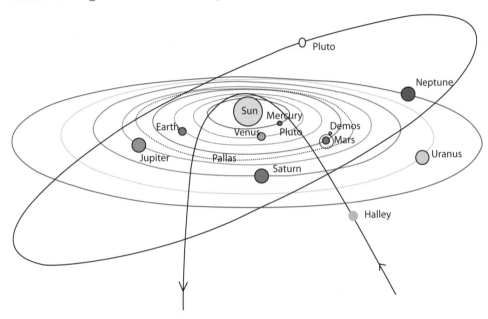

Complete this table, using the diagram above and these labels.

| asteroid | comet | dwarf planet | moon | planet | star |

Type of object	Description	Example
	large massive luminous object at very high temperature	
	one of a number of small objects orbiting between Mars and Jupiter	
	object orbiting a planet	
	object with a very elliptical orbit that comes from outside the solar system and passes close to the Sun	
	large massive object orbiting a star	
	smaller object orbiting a star	

Supernovae

Draw a ring around the correct words in **bold** to complete the statement.

A supernova is what we see when a massive **star** / **planet** explodes. One particular type of supernova always gives the same peak **size** / **luminosity**. This means we can use their **brightness** / **size** to measure the distance to **stars** / **galaxies** far away from the Milky Way, in a similar way to the way Cepheid variable stars are used to measure **distance** / **age**.

E Observing distant galaxies

1 Redshift

a Draw a (ring) around the correct **bold** words to complete this statement.

Hubble discovered that the wavelengths and frequencies of electromagnetic radiation arriving from distant galaxies were **redshifted / blueshifted**. This means that the wavelengths were **longer / shorter**.

b Draw one line from each wavelength change to show whether it is a redshift or a blueshift.

Wavelength change   **This change is a . . .**

| microwave range to red range |

| red range to blue range |

| blue range to green range | | redshift |

| X-ray range to violet range |

| red range to microwave range | | blueshift |

| 589 nm to 610 nm |

| 456 nm to 337 nm |

2 Moving galaxies

Use words from the box to complete these sentences.

| speed of recession   spectra   Hubble   away from   Cepheid variable |

By analysing the .... of .... stars in 46 galaxies,

.... discovered that all distant galaxies are moving .... ours.

He also found that a galaxy's .... is proportional to its distance

away from us.

The Hubble equation

Speed of recession = Hubble constant × distance

Complete the table showing data for different galaxies, using the Hubble equation.

Speed of recession	Hubble constant	Distance
5000 km/s	70 km/s per Mpc	.... Mpc
3500 km/s	.... km/s per Mpc	48 Mpc
.... km/s	2.3×10^{-18} s^{-1}	3.08×10^{21} km
2000 km/s	2.3×10^{-18} s^{-1}	.... km
3000 km/s	.... s^{-1}	1.23×10^{21} km

The expanding Universe

This diagram shows a wave before and after it has been redshifted to a longer wavelength.

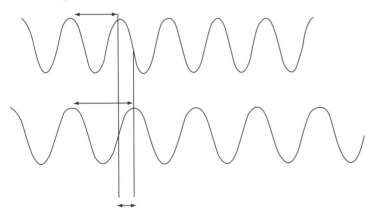

a On the diagram, label:

- the original wavelength
- the redshifted wavelength
- the redshift

b Tick ✓ the correct phrase to complete this statement.

Most scientists think the explanation for this behaviour is that:

☐ the galaxies are all drifting through space in random directions

☐ the galaxies are all travelling very fast through space

☐ the space between the galaxies is expanding

☐ the speed of light is not constant in deep space

☐ the wavelengths corresponding to the colours are different in other galaxies

c This graph shows the speed of recession of different galaxies plotted against their distance away from Earth.

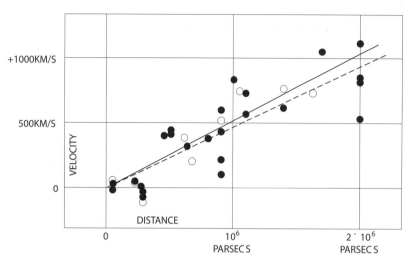

Use the graph to complete these statements.

i A galaxy 10^4 parsecs away is moving away with a velocity of about _____ km/s.

ii A galaxy 2×10^4 parsecs away is moving away with a velocity of about _____ km/s.

iii The data points are scattered about because _____

_____

iv A best-fit straight line though the scattered points passes through (0, 0). This shows

that _____

_____

⑤ The big bang theory

Hubble's graph suggested that the Universe was expanding. As a result scientists came up with the big bang theory.

a Complete this statement.

The big bang theory is a theory to explain _____

_____

b Describe the big bang theory.

_____

_____

c Some scientists did not agree with the big bang theory when it was first suggested.

Give two reasons why some scientists might not have agreed with the new theory.

1 _____

2 _____

Topic 4: Stars

How hot is the Sun?

The spectrum of a star

The graph shows the spectra produced by three different stars.

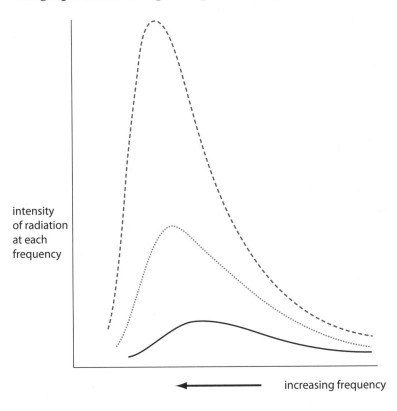

intensity
of radiation
at each
frequency

← increasing frequency

a Label the hottest star and the coolest star.

b Explain how astronomers analyse starlight to work out a star's temperature.
Use the words in the box to help you.

electromagnetic radiation	frequencies	temperature	spectrometer
peak frequency	telescope	intensity of radiation at each frequency	

....

....

....

....

....

....

① The composition of stars

The diagram shows the dark lines seen in the spectrum of visible light from a star.

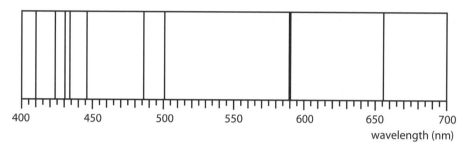

a Use the table below to identify the elements present in this star. Put a tick ✓ or a cross ✗ in every box in the last column.

Element	Wavelengths (nm)	Present in the star?
calcium	423, 431	
helium	447, 502, 588	
hydrogen	410, 434, 486, 656	
iron	431, 438, 467, 496, 527	
sodium	589, 590	

b Complete these statements.

When the spectrum of the Sun was first observed the dark lines at 447 nm,

502 nm, and .... could not be identified.

The scientists predicted that they were .... lines of an

.... that had not yet been discovered on ....

They named it ....

Energy levels and line spectra

a Use words from the list to complete the following sentences.

atom	atoms	electron	energy	level
levels	photon	photons		

Atoms of a particular element all have the same allowed ....

Emission spectra are formed because the atoms give out a ....

of light when an .... drops to a lower ....

This happens because the .... loses energy. The difference in

energy between the two .... is equal to the energy of

the emitted ....

Absorption spectra are formed because a .... of light is

absorbed by an .... when it moves to a higher ....

.... in the atom.

Line spectra can be used to identify elements because the frequencies of

the .... match the .... of the

.... and these are different for each element.

b The diagram shows possible energy levels in an atom.

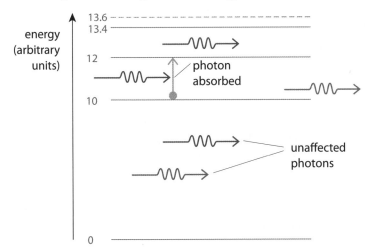

i In the diagram, the photon that is absorbed has .... energy units.

ii Explain whether a photon with 1.4 energy units could be absorbed by an electron.

....

iii Explain whether a photon with 1 energy unit could be absorbed by an electron.

....

135

c Explain why this element emits light that makes a line spectrum rather than a continuous spectrum.

....

....

....

....

d The dotted line marked 13.6 represents the energy needed for ionisation. An electron with this amount of energy, or more, would be able to escape from the atom. Use this mechanism to explain the difference between ionising and non-ionising radiation.

....

....

....

....

C   What fuels the Sun?

① The nucleus

a What two particles are found in the nuclei of atoms?

....

b Particles found in the nuclei of atoms are called nucleons. How does the mass of a nucleon compare with the mass of an electron?

....

c In a stable nucleus, two forces are balanced: the force that holds nuclear particles together and the force that tries to push some of them apart are equal. Describe the two forces by completing the table.

	Name of force	Particles that the force acts on	Range of the force
Force holding particles together in nucleus			
Force pushing particles apart in nucleus			

Nuclear fusion

This explanation solved the mystery of the source of the Sun's energy.

For nuclear fusion to occur, two nuclei must overcome their **repulsion** and get close enough for the **attractive force** to make them join together and make a new nucleus with a larger mass. The nuclei have **kinetic energy** before and after a fusion reaction. The process of nuclear fusion releases energy, so the total kinetic energy after a reaction is greater. Fusion takes place in the core of a star because of **conditions** found there.

Explain each of the terms in **bold**.

repulsion: ....

....

attractive force: ....

....

kinetic energy: ....

....

conditions: ....

....

....

Hydrogen fusion

Inside stars like the Sun, hydrogen fusion takes place.

a Draw a ring around the correct **bold** words to complete this description of the reaction.

When two hydrogen **atoms / nuclei** fuse they produce a **deuterium / helium** nucleus and a small positively charged particle, similar to an electron, called a **positron / proton**.

The deuterium **atom / nucleus** then combines with another **helium / hydrogen** nucleus to produce a **helium-3 / hydrogen-3** nucleus.

This **helium-3 / hydrogen-3** nucleus then combines with another one that has been produced in the same way. This reaction produces a **helium-4 / hydrogen-4** nucleus plus two **deuterium / hydrogen** nuclei.

b Complete the nuclear equations for the reactions:

$$^{1}_{1}\text{H} + ^{1}_{1}\text{H} \longrightarrow \underline{\quad} + ^{0}_{+1}\text{e}^{+}$$

$$^{1}_{\underline{\quad}}\text{H} + ^{2}_{\underline{\quad}}\text{H} \longrightarrow \underline{\quad}$$

$$^{3}_{\underline{\quad}}\text{He} + ^{\underline{\quad}}_{2}\underline{\quad} \longrightarrow \underline{\quad} + 2\ ^{1}_{\underline{\quad}}\text{H}$$

These can be summarised by the reaction:

$$4\ ^{1}_{1}\text{H} \longrightarrow ^{4}_{2}\text{He} + 2\ ^{0}_{+1}\text{e}^{+}$$

④ **Energy from fusion**

When nuclei fuse to give a nucleus with mass less than the iron-56 nucleus mass is lost and energy is released according to Einstein's equation.

$E = mc^2$

a Complete these statements about the equation.

In the equation $E = mc^2$

i E is .... transferred measured in units ....

ii m is the difference in .... between ....

.... measured in units ....

iii c is .... measured in units ....

b 0.5 g of mass is converted to energy of the products in fusion reactions. Calculate the energy released.

(speed of light $= 3 \times 10^8$ m/s)

....

....

....

What are other stars like?

1 Stars change

Use words from the box to complete these sentences.

data	colour	H–R diagram	luminosity	white dwarf
red giants	models	main-sequence	small part	

Astronomers observe stars of quite different _____ and _____ .

When stars are plotted on a _____ (a chart of luminosity against

temperature), they fall into three main groups: main-sequence stars, _____

stars, and _____ or supergiants. Linking _____ about

star populations to _____ of how stars work, astronomers conclude

that stars change, and that:

An average star spends most of its lifetime as a _____ star.

A star may spend a _____ of its lifetime as a red giant or as a white dwarf.

2 The Hertzsprung–Russell diagram

The H–R diagram is a graph of luminosity of stars plotted against temperature.

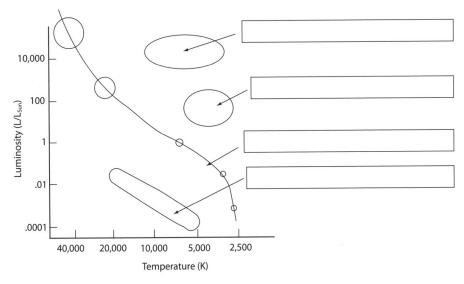

a Draw a labelled vertical arrow at the side of the y-axis going from 'faint' to 'luminous'.

b Draw a labelled horizontal arrow along the x-axis going from 'cool' to 'hot'.

c Use words from this list to label the different groups of star on the diagram.

giants	main sequence	supergiants	white dwarfs

d Mark the position of our Sun on the diagram with an X.

① **The behaviour of gases**

a What four quantities are needed to fully describe the properties of a sample of a gas?

....

b The kinetic model of matter says that all matter consists of tiny particles (often molecules) in motion. The following statements explain how a gas exerts pressure on the walls of its container. Draw lines to match the two parts of each statement.

The billions of molecules in a gas . . .	. . . is related to the temperature of the gas.
The speed of the molecules . . .	. . . causes a tiny force.
As the molecules move around, they . . .	. . . move around freely in what is mostly empty space.
Each collision with the walls . . .	. . . together produce gas pressure on the walls.
The tiny forces from molecular collisions with the walls . . .	. . . collide with each other and with the walls of their container.

c Use the model of molecular collisions to explain:

i why the pressure of a gas increases when the volume of a gas is reduced, with its temperature constant.

....

....

ii why the pressure of a gas increases with temperature, when its volume stays constant.

....

....

iii why the volume of a gas increases with temperature, when its pressure stays constant.

....

....

iv what a temperature of 'absolute zero' means.

....

....

Temperature scales

a Complete this sentence.

The absolute zero of temperature occurs at _____ °C or _____ K.

b Complete the table by converting each temperature from one scale to the other.

At this temperature ...	Temperature (K)	Temperature (°C)
iron melts		1540
aluminium melts	933	
		100
human body	310	
typical room		20
carbon dioxide freezes	195	
nitrogen boils		−196
helium boils	4	

The gas laws

The pressure, temperature, and volume of a fixed mass of gas follow these relationships.

pressure × volume = constant at constant temperature or $p_1 V_1 = p_2 V_2$

$\dfrac{\text{pressure}}{\text{temperature}}$ = constant at constant volume or $\dfrac{p_1}{T_1} = \dfrac{p_2}{T_2}$

$\dfrac{\text{volume}}{\text{temperature}}$ = constant at constant pressure or $\dfrac{V_1}{T_1} = \dfrac{V_2}{T_2}$

a Draw a ring around the correct **bold** words to complete these statements.

i For the relationships between pressure or volume and temperature to be true the temperature must be the **absolute** / **average** temperature measured in **Celsius** / **Kelvin**.

ii When the pressure on a gas doubles, and the temperature does not change, the volume **doubles** / **halves**.

iii When a gas expands to fill four times the volume, and the temperature does not change, the pressure is **one quarter** / **four times** the original value.

iv A container holds gas at 2°C. It does not expand when heated. The pressure on it has doubled when the temperature reaches **4°C** / **550°C** / **277°C**.

b Choose a sketch graph below that is the correct shape for each table entry, and write the labels on the axes. There is one 'odd one out' graph that does not represent a relationship in the table.

Vertical (**y**) axis	Horizontal (**x**) axis
pressure (Pa)	temperature (°C)
pressure (Pa)	temperature (K)
pressure (Pa)	volume (m³)
volume (m³)	temperature (°C)
volume (m³)	temperature (K)

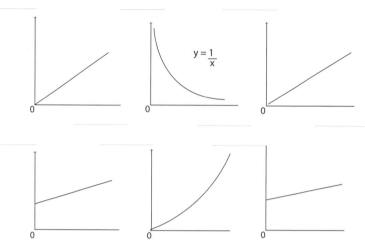

c Calculate the final volume of 2 m³ of air at 27 °C heated at constant pressure until its temperature is 627 °C.

convert temperatures to Kelvin

$T_1 = 27\,°C =$ .... K $T_2 = 627\,°C =$ .... K

$V_1 = 2\,m^3$ so

$$\frac{\rule{4cm}{0.4pt}}{\rule{4cm}{0.4pt}} = \frac{V_2}{\rule{4cm}{0.4pt}}$$

$V_2 =$ .... m^3

d Calculate the final temperature of a cylinder of gas that does not expand when heated from 27°C. The pressure increased from 102 kPa to 255 kPa.

convert temperature to Kelvin

$T_1 = 27\,°C =$ .... K

$p_1 =$ .... kPa $p_2 =$ .... kPa

$$\frac{\rule{4cm}{0.4pt}}{\rule{4cm}{0.4pt}} = \frac{\rule{4cm}{0.4pt}}{T_2}$$

$T_2 =$ .... K

Convert to celsius $T_2 =$ .... °C

How do stars form?

Protostars

a Number the following statements in order, so that they describe the formation of a star with its solar system.

	Material further out in the disc clumps together to form planets.

	Eventually the temperature at the centre is hot enough for fusion reactions to occur and a star is born.

	A cloud of dust and hydrogen in space starts to contract, pulled together by gravity. It becomes a rotating disc.

	The temperature increases when the raw material is compressed, getting hotter and hotter at the centre.

b Use the kinetic model of matter to explain why the temperature in the centre of a protostar rises. Include in your explanation the role played by gravity.

c Explain why gravity gives any large mass of material a spherical shape.

② Main-sequence stars

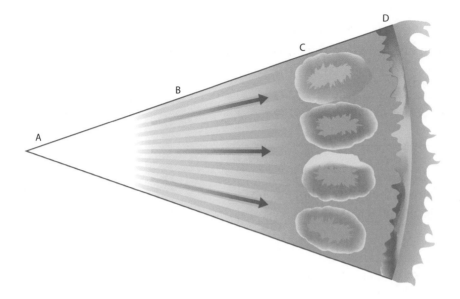

a The diagram shows the internal structure of a main-sequence star.

Draw lines to match the labels and the descriptions of what happens in each part of the star.

Label	Part of the star	What happens there
A	photosphere	fusion of hydrogen to form helium takes place
B	radiative zone	energy is transferred by convection cells
C	core	energy radiates into space
D	convective zone	energy is carried by photons

b Fill in the blanks to complete the sentence.

How long a star lasts in its main-sequence phase depends on its ....

and ....

c Complete these statements.

i In all the main-sequence stars hydrogen ....

....

ii The most massive stars spend the shortest time on the main sequence because

....

Main-sequence stars and the H–R diagram

These are the axes for a Hertzsprung–Russell diagram.

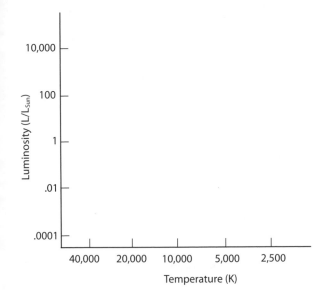

a On the graph above, sketch a line showing where the main-sequence stars are.

b On the graph add these labels to the ends of the main sequence:

- most luminous main-sequence stars

- faintest main-sequence stars

- most massive main-sequence stars

c Mark the position of the Sun on the main sequence.

How do stars end?

The end of a star

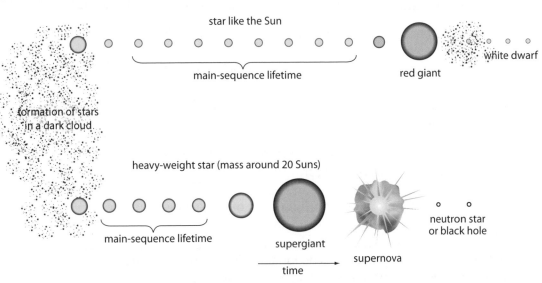

a Describe similarities and differences between the life cycles of stars like the Sun, and stars of much greater mass.

Similarities in their life cycles: ....

....

Differences in their life cycles: ....

....

b Use words from the box to complete these sentences.

kinetic	helium	decreases	gravitational	oxygen
red giant	hydrogen	carbon	increases	

As the .... of a main-sequence star runs out, its core cools

down and so its volume .... The collapse transfers ....

energy to .... energy of helium nuclei, which means the star's

core temperature .... This restarts the fusion process,

with .... changing to nuclei of even bigger mass such as .... ,

nitrogen, and .... The energy this releases produces a ....

or supergiant.

② **The final stages**

a Fill in the blank to complete these sentences.

When a red giant runs out of helium, its mass is too small for gravity to compress its

core and produce higher temperatures, and so fusion stops. The star shrinks into

a hot .... , which gradually cools.

b The following statements describe what happens to supergiants. Number the statements in the correct order.

☐	The supernova remnant becomes either a neutron star or a black hole. Remnants with the biggest masses become black holes.
☐	When eventually the star produces iron, it runs out of nuclear fuel. The rate of fusion in the core decreases and pressure falls.
☐	The weight of outer layers of the star is no longer balanced by the core's pressure. The star dramatically collapses and then explodes as a supernova.
☐	Fusion in a supergiant continues to produce heavier and heavier elements, because gravity causes such high pressures in the core of massive stars.

Are we alone?

Exoplanets

a Draw a (ring) around the approximate number of galaxies in the Universe.

| millions   tens of millions   hundreds of millions   thousands of millions |

b Draw a (ring) around the approximate number of stars in the Milky Way.

| 1 000 000   100 000 000   100 000 000 000   1 000 000 000 000 000 |

c Draw a (ring) around the approximate number of exoplanets discovered so far.

| 50   500   5 000   5 000 000 |

d Explain how exoplanets are detected.

....

....

e Suggest a reason why not many exoplanets have been observed and why those discovered are all more massive than Earth.

....

f Give a reason why many scientists think it is likely that life exists elsewhere in the Universe.

....

SETI

SETI stands for the Search for Extra-Terrestrial Intelligence.
Tick ✓ the correct statements.

☐ Some signs of extraterrestrial life on Mars have been discovered.

☐ No sign of extraterrestrial life has ever been discovered.

☐ No sign of intelligent extraterrestrial life has ever been discovered.

☐ Some signs of extraterrestrial life have been discovered in meteorites.

Appendices

Useful relationships, units, and data
Relationships
You will need to be able to carry out calculations using these mathematical relationships.

B7 Further biology

$$BMI = \frac{body\ mass\ (kg)}{[height\ (m)]^2}$$

C7 Further chemistry

$$concentration\ of\ a\ solution = \frac{mass\ of\ solute}{volume\ of\ solution}$$

$$percentage\ yield = \frac{actual\ yield}{theoretical\ yield} \times 100\%$$

chromatography: retardation factor (R_f) =

$$\frac{distance\ travelled\ by\ solute}{distance\ travelled\ by\ solvent}$$

P7 Studying the Universe

$$power\ of\ a\ lens = \frac{1}{focal\ length}$$

magnification of a telescope =

$$\frac{focal\ length\ of\ objective\ lens}{focal\ length\ of\ eyepiece\ lens}$$

Hubble equation: speed of recession = Hubble constant × distance

Einstein's equation: $E = mc^2$ where E is the energy produced, m is the mass lost and c is the speed of light in a vacuum

For a fixed mass of gas:

pressure × volume = constant *at constant temperature*

$$\frac{pressure}{temperature} = constant\ for\ constant\ volume$$

$$\frac{volume}{temperature} = constant\ at\ constant\ pressure$$

Units that might be used in the Physics course

length: metres (m), kilometres (km), centimetres (cm), millimetres (mm), micrometres (μm), nanometres (nm)

mass: kilograms (kg), grams (g), milligrams (mg)

time: seconds (s), milliseconds (ms)

temperature: degrees Celsius (°C); Kelvin (K)

area: cm^2, m^2

volume: cm^3, dm^3, m^3, litres (l), millilitres (ml)

speed and velocity: m/s, km/s, km/h

force: newtons (N)

energy/work: joules (J), kilojoules (kJ), megajoule (MJ), kilowatt-hours (kWh), megawatt-hours (MW

distance (astronomy): parsecs (pc)

power of a lens: dioptres (D)

Prefixes for units

nano	micro	milli	kilo	mega	giga	tera
one thousand millionth	one millionth	one thousandth	× thousand	× million	× thousand million	× million million
0.000000001	0.000001	0.001	1000	1000 000	1000 000 000	1000 000 000 000
10^{-9}	10^{-6}	10^{-3}	$\times 10^3$	$\times 10^6$	$\times 10^9$	$\times 10^{12}$

Useful Data

P7 Studying the Universe

solar day = 24 hours

sidereal day = 23 hours 56 minutes

age of the Universe: approximately 14 thousand million years

absolute zero of temperature , $0\,K = -273\,°C$

Chemical Formulae

C7 Further Chemistry

methanol CH_3OH, ethanol C_2H_5OH

methanoic acid $HCOOH$, ethanoic acid CH_3COOH

The Periodic Table

Key:

proton number	1
symbol	H
name	hydrogen
atomic mass	1

(relative atomic mass shown above symbol; proton number shown below name)

group number →

Period	1	2	transition metals										3	4	5	6	7	0
1	1 H hydrogen 1																	4 He helium 2
2	7 Li lithium 3	9 Be beryllium 4											11 B boron 5	12 C carbon 6	14 N nitrogen 7	16 O oxygen 8	19 F fluorine 9	20 Ne neon 10
3	23 Na sodium 11	24 Mg magnesium 12											27 Al aluminium 13	28 Si silicon 14	31 P phosphorus 15	32 S sulfur 16	35.5 Cl chlorin 17	40 Ar argon 18
4	39 K potassium 19	40 Ca calcium 20	45 Sc scandium 21	48 Ti titanium 22	51 V vanadium 23	52 Cr chromium 24	55 Mn manganese 25	56 Fe iron 26	59 Co cobalt 27	59 Ni nickel 28	63.5 Cu copper 29	65 Zn zinc 30	70 Ga gallium 31	73 Ge germanium 32	75 As arsenic 33	79 Se selenium 34	80 Br bromine 35	84 Kr krypton 36
5	86 Rb rubidium 37	88 Sr strontium 38	89 Y yttrium 39	91 Zr zirconium 40	93 Nb niobium 41	96 Mo molybdenum 42	98 Tc technetium 43	101 Ru ruthenium 44	103 Rh rhodium 45	106 Pd palladium 46	108 Ag silver 47	112 Cd cadmium 48	115 In indium 49	119 Sn tin 50	122 Sb antimony 51	126 Te tellurium 52	127 I iodine 53	131 Xe xenon 54
6	133 Cs caesium 55	137 Ba barium 56	139 La lanthanum 57	178 Hf hafnium 72	181 Ta tantalum 73	184 W tungsten 74	186 Re rhenium 75	190 Os osmium 76	192 Ir iridium 77	195 Pt platinum 78	197 Au gold 79	201 Hg mercury 80	204 Tl thallium 81	207 Pb lead 82	209 Bi bismuth 83	209 Po polonium 84	210 At astatine 85	222 Rn radon 86
7	223 Fr francium 87	226 Ra radium 88	227 Ac actinium 89	104	105	106	107	108	109	110	111	112						

UNIVERSITY PRESS

Great Clarendon Street, Oxford OX2 6DP

Oxford University Press is a department of the University of Oxford.
It furthers the University's objective of excellence in research,
scholarship, and education by publishing worldwide in

Oxford   New York

Auckland   Cape Town   Dar es Salaam   Hong Kong   Karachi
Kuala Lumpur   Madrid   Melbourne   Mexico City   Nairobi
New Delhi   Shanghai   Taipei   Toronto

With offices in

Argentina   Austria   Brazil   Chile   Czech Republic   France   Greece
Guatemala   Hungary   Italy   Japan   Poland   Portugal   Singapore
South Korea   Switzerland   Thailand   Turkey   Ukraine   Vietnam

Oxford is a registered trade mark of Oxford University Press
in the UK and in certain other countries.

British Library Cataloguing in Publication Data.

Data available.

ISBN 978-0-19-913851-7

10 9 8 7 6 5 4 3 2

Printed in Great Britain by Bell and Bain Ltd, Glasgow.

Paper used in the production of this book is a natural, recyclable product made
from wood grown in sustainable forests. The manufacturing process conforms to
the environmental regulations of the country of origin.

Acknowledgements

Illustrations by IFA Design, Plymouth, UK, Clive Goodyer, and Q2A Media.

Project Team acknowledgements
These resources have been developed to support teachers and students
undertaking the OCR suite of specifications GCSE Science Twenty First Century
Science. They have been developed from the 2006 edition of the resources.
We would like to thank David Curnow and Alistair Moore and the examining
team at OCR, who produced the specifications for the Twenty First Century
Science course.

Authors and editors of the first edition
We thank the authors and editors of the first edition, Jenifer Burden, Peter
Campbell, Andrew Hunt, Robin Millar, and Caroline Shearer.

Many people from schools, colleges, universities, industry, and the professions
contributed to the production of the first edition of these resources. We also
acknowledge the invaluable contribution of the teachers and students in the
pilot centres.

The first edition of Twenty First Century Science was developed with support
from the Nuffield Foundation, The Salters Institute, and the Wellcome Trust.

A full list of contributors can be found in the Teacher and Technician
Resources.

The continued development of Twenty First Century Science is made possible
by generous support from:
- The Nuffield Foundation
- The Salters' Institute